FOOD SCIENCE AND TECHNOLOGY

THE DIVERSIFIED BENEFITS OF COCOA AND CHOCOLATE

FOOD SCIENCE AND TECHNOLOGY

Additional books in this series can be found on Nova's website under the Series tab.

Additional e-books in this series can be found on Nova's website under the eBooks tab.

FOOD SCIENCE AND TECHNOLOGY

THE DIVERSIFIED BENEFITS OF COCOA AND CHOCOLATE

BONIFACIA ZAYAS ESPINAL
EDITOR

nova
science publishers
New York

Library of Congress Cataloging-in-Publication Data

ISBN: 978-1-53613-258-8

Published by Nova Science Publishers, Inc. † New York

CONTENTS

PREFACE

The Diversified Benefits of Cocoa and Chocolate presents the different available methodologies for analysis, quantification, isolation, purification, and structure elucidation of polyphenols in cocoa and cocoa-derived products are reviewed. Taking into account that there is evidence that demonstrates that the bioactivity of flavanols is significantly influenced by their stereochemical configuration, enantioselective methods have also been included such as chiral capillary electrophoresis, micellar electrokinetic chromatography and ultra-performance liquid chromatography. The authors state that about forty million people die annually as a consequence of noncommunicable diseases accounting for approximately 70% of all death globally. Therefore, the current research concerning the benefits of dark chocolate regarding noncommunicable diseases and the associated risk factors is examined. Following this, the authors introduce chocolate as a cocoa carrier of probiotics and bioactive whey protein hydrolysate. This research examined the influence of additional ingredients on the functional properties and rheology of the final product. Another study is included with the objective of exploring the potential of NIR usage for monitoring the cocoa powder and sugar mixtures composition. Based on the obtained results, NIRs analysis in combination with the PCA and the PLS has proved to be an adequate, fast and non-invasive method for the cocoa powder drink mix composition prediction. A separate study attempts to confirm the relationship between increased pollinator abundance and higher yields of cocoa pods at one locality in Costa Rica. While cocoa pollination can depend upon a diverse array of ceratopogonid midges, in this study, one species *Forcipomyia youngii*, dominated the samples. Research is included with the aim of assessing the proteins contents, digestibility, and amino acid profile of the milled cocoa bean husks from the roasted cocoa, and to propose it as an ingredient for PKU food formulations. The authors determine that Cocoa can be used as a raw material in products destined for special regimes, after verification of its microbiological safety. The final study examined and compared the content of total polyphenols, minerals (potassium, calcium, magnesium, iron, zinc, copper, manganese), total dietary fiber and soluble dietary fiber in in different chocolate products collected from the market.

Chapter 1 - Cocoa (*Theobroma cacao*), from the health viewpoint, is an important commodity in the world due to the high contents of polyphenols which are its main antioxidant-active fraction, particularly flavonoids. Polyphenols also confer astringent and bitter taste, also contributing to the fruity and green flavors of cocoa liquors. There are three main groups of flavonoids in cocoa: catechins (flavan-3-ols), anthocyanins, and proanthocyanidins. Other minor polyphenols also present in cocoa include flavones, phenolic acids, clovamide, dideoxyclovamide, resveratrol, and piceid. The content and composition of polyphenols in cocoa and its derived products vary strongly depending on genotype, cultivar, origin, agricultural practices, post-harvest practices, and bean processing. Fermentation of cocoa beans is considered to be one of the major steps that affect polyphenol content. Anthocyanins usually disappear rapidly during this process being hydrolyzed to anthocyanidins, which polymerize along with simple catechins to form complex tannins. During the drying process, the oxidation reactions that started in fermentation continue, the phenol degradation increasing with temperature. Alkalinization of the nibs, liquor, or powder also originates considerable losses of cocoa polyphenols. Therefore, given the functional effects of the polyphenols above mentioned, it is vital analyzing the total phenolic content and the composition and quantity of individual polyphenols in both cocoa beans and cocoa-derived products, being chocolate the most representative of these latter. The quantitative determination of total polyphenols in cocoa and chocolate usually includes the assessment of the total phenol (TPC) or flavonoid (TFC) contents with *Folin-Ciocalteu* reagent, the measurement of the flavan-3-ols content by the vanillin and *p*-dimethylaminocin-namaldehyde (*p*-DAC) assays or the determination of proanthocyanidins by the butanol/HCl assay. The analysis of the polyphenols profile has usually been accomplished by liquid chromatography (HPLC) with different detectors. Nevertheless, nowadays liquid chromatography-mass spectrometry (HPLC-MS) or liquid chromatography-tandem mass spectrometry (HPLC-MS/MS) are among the most widely used techniques for both structural characterization and quantification of both low and high molecular weight polyphenols in cocoa. In the last years, there have been significant improvements in HPLC technique with the development of ultra-performance liquid chromatography (UPLC) and today, the number of studies focusing on the analysis of polyphenols by using UPLC coupled to MS (UPLC-MS/MS) is increasing. In this chapter, the different current available methodologies for analysis, quantification, isolation, purification, and structure elucidation of polyphenols in cocoa and cocoa-derived products are reviewed. Taking into account that there is evidence that demonstrates that the bioactivity of flavanols is significantly influenced by their stereochemical configuration, enantioselective methods have also been included such as chiral capillary electrophoresis (CCE), micellar electrokinetic chromatography (MEKC) or UPLC.

Chapter 2 - About forty million people die annually as a consequence of noncommunicable diseases (NCDs), accounting for approximately 70% of all death globally. Cardiovascular diseases account for 17.7 million deaths annually, followed by cancers (8.8 million), respiratory diseases (3.9 million) and diabetes mellitus, which is responsible for 1.6 million deaths. Smoking, physical inactivity, unhealthy diet (eating out and consuming take-away foods, which are high in salt and low in fibre) and excessive alcohol intake have all been identified as established risk factors for NCDs. The preventive strategies for NCDs have been focused on these modifiable risk factors. The recommendation of healthy dietary practices such as eating a healthy breakfast, preparing home-made meals low in salt and fat, and eating sufficient quantities of fruits and vegetables, is one strategy used by health professionals to curb NCDs. Fruits and vegetables are rich in dietary fibre and also a good source of vitamins, minerals and (poly)phenols. Epidemiological studies have shown that the consumption of fruits and vegetables is negatively correlated with the incidence of NCDs. However, other foods such as dark chocolate contain higher concentrations of (poly)phenols per 100 g compared with fruits and vegetables. Dark chocolate contains an average of 1860 mg total (poly)phenols per 100 g fresh weight. Numerous studies have shown the diverse health benefits of dark chocolate, particularly against NCDs, which are due to its high (poly)phenol content. Hence, this chapter will examine the current research in relation to the benefits of dark chocolate vis-à-vis NCDs and the associated risk factors.

Chapter 3 - The aim of this chapter was to introduce chocolate as a cocoa carrier of probiotics and bioactive whey protein hydrolysate. This research examined the influence of additional ingredients on the functional properties and rheology of the final product. The chocolate prepared with *L. rhamnosus* HN001 strain at 40°C achieved the highest survivability, which was above 90% (viable cell count of about 8.1 log (CFU/g)) after 6 months of storage. This strain could be selected for high-quality probiotic chocolate production. The addition of probiotic *L. rhamnosus* HN001 significantly improves the rheological properties of probiotic chocolate. The addition of bioactive whey protein hydrolysate did not show significant changes in the total content of polyphenols or ABTS radical scavenging activity. However, DPPH radical scavenging activity showed an increase of 1.28 times during storage and about 1.46 in relation to the control sample and the sample supplemented with unhydrolyzed whey proteins. The bioactive whey protein hydrolysate affects chocolate rheology by increasing the thixotropic loop 8.02 times in relation to the control sample, and 1.57 times compared to the sample supplemented with unhydrolyzed whey proteins. Based on the results, the addition of probiotics and bioactive whey protein hydrolysate powder in industrial conditions, which was done in the mixing tank (at 40°C) before the chocolate shaping phase, provides preservation of both functional ingredients as key components for this type of product. The addition of probiotics improves the rheological properties of chocolate, while the addition of whey protein hydrolysate leads to undesirable viscosity increase that can be avoided using a more efficient emulsifier.

Chapter 4 - Tropical tree *Theobroma cacao* L. is a source of one of the most highly appreciated raw materials in the food industry: cocoa beans. Raw cocoa beans are used for the production of chocolate, cocoa mass, cocoa butter and cocoa powder. Cocoa powder is a product which is usually lower priced in comparison to cocoa butter and chocolate, but it is further used in many more foodstuffs as a raw material. One of the widely used product which includes cocoa powder as a raw material is cocoa powder drink mix. It is produced by mixing cocoa powder with sugar, usually in a 30:70% ratio. Besides cocoa and sugar, milk powder and instantizers (soy lecithin) are added to the mixtures prior to agglomeration and drying. In order to achieve and maintain a constant mixture composition, it is of great importance to use adequate mixing equipment, as well as to monitor the mixing process at all stages. Since industry mixing devices contain large tonnages of powdered material, constant sampling and mixture composition analysis can be somewhat difficult and time consuming. Because of that, the need for a fast, in-line, build-in mixture composition monitoring technique arises. Near infrared spectroscopy (NIRs) is a valuable analytical tool in agriculture and the food industry for rapid and accurate analysis of certain product constituents. In addition to being rapid, NIRs requires very little sample preparation, requires no chemicals, and produces no chemical waste. In this work, the objectives were to explore the potential of NIR usage for monitoring the cocoa powder and sugar mixtures composition. Cocoa powder and powdered sugar mixtures were produced in different ratios (0 – 100%) with the aim of creating a calibration curve based on the NIR spectra recorded for each mixture. The recorded NIR spectra were then subjected to the Principal Component Analysis (PCA) to obtain the factor values of principal components. The factor values were then plotted in relation to the mixture composition to obtain a calibration curve. The

Partial Least Squares regression (PLS) was performed to obtain and validate models based on which the concentration of sugar at any given moment in the mixture can be quantitatively and/or qualitatively predicted. Based on the obtained results, NIRs analysis in combination with the PCA and the PLS has proved to be an adequate, fast and non-invasive method for the cocoa powder drink mix composition prediction.

Chapter 5 - The world's production of cocoa beans is directly dependent on the pollination of cocoa flowers by insects, chiefly midges of the diptern family Ceratopogonidae and to a lesser extent, Cecidomyiidae. Previous studies in Central America suggested that increases in cocoa pod production is directly related to increased abundance of these pollinators. In these studies, banana pseudostem pieces placed in piles at the base of cocoa trees, and upon rotting under humid conditions, provided ideal breeding sites for the life cycle of cocoa-pollinating midges. The study presented in this paper is an attempt to confirm the relationship between increased pollinator abundance and higher yields of cocoa pods at one locality in Costa Rica. While cocoa pollination can depend upon a diverse array of ceratopogonid midges, in this study, one species *Forcipomyia youngii*, dominated the samples. The technique used was a four-year "cross over" study in which two plots, each with 25 cocoa trees under a shade canopy of chiefly *Cordia alliodora* "laurel" trees received a treatment of pseudostems twice. If pod production significantly increased with the treatment, then the treatment was effective when applied to Lot A over two years and when applied to Lot B for another two years. For each plot when given the treatment, an increased abundance of pollinators would result in increased pollination and more pods. Analysis of the data revealed no significant increase in pod production with treatment but trends in that direction. Yet the authors conclude that this technique could demonstrate increases in cocoa pod production under different environmental conditions. While the technique is less effective in highly humid conditions as the present study site, it could be far more effective as a means of expanding cocoa growing in drier regions and perhaps in association with water-drip irrigation. Under dry conditions, moist and humid piles of banana pseudostems would provide a suitable breeding niche for pollinators. Under these conditions, a crossover in breeding substrate treatment for pollinating midges is predicted to be far more effective than under uniformly humid conditions. Thus, there is a potential agronomic benefit is in the application of this technique. It can diversify where cocoa trees are grown, i.e., in drier regions. Therefore, this study warrants replication along an environmental gradient, from very humid to very dry conditions. The use of banana pseudostems to breed pollinating midges is a relatively inexpensive technique using a resource (banana plants) typically available where cocoa is grown. These findings are further discussed in terms of the environmental benefits of growing cocoa as a cash crop integrated into agroforestry systems.

Chapter 6 - The raw materials used for chocolate production provide a source of proteins, carbohydrates, fats, minerals, and vitamins, which are essential for growth and development. Dark chocolate is also rich in polyphenolic compounds, originating from dark solids of cocoa beans. Milk chocolate contains less cocoa bean solids and thus less polyphenolic compounds comparing to dark chocolate. On the other hand, white chocolate contains only cocoa butter and thus lack bioactive components that have a positive impact on human health. Cocoa bean contains several minerals, some of which are found in high amounts in processed chocolate. The amount of retained minerals depends on the content of cocoa bean solids in chocolate. Accordingly, dark chocolate typically has a higher amount of minerals than milk or white chocolate. It is also known that unprocessed cocoa bean presents a good source of dietary fiber, mostly insoluble fiber, which is significantly reduced when removing the cocoa husk. Dark chocolate does not contribute significantly to dietary fiber intake, but, however, contains a higher amount of dietary fiber compared to milk chocolate. This research examined and compared the content of total polyphenols, minerals (potassium, calcium, magnesium, iron, zinc, copper, manganese), total dietary fiber and soluble dietary fiber in in different chocolate products collected from the market: white chocolate (25% of cocoa solids – cocoa butter), milk chocolate (25% of cocoa solids), baking chocolate containing 44% of cocoa solids, and dark chocolates containing 58%, 75%, 88%, and 100% of cocoa solids.

In: The Diversified Benefits of Cocoa … ISBN: 978-1-53613-258-8
Editor: Bonifacia Zayas Espinal

Chapter 1

Analytical Methodologies for the Assessment of Polyphenols in Cocoa and Cocoa Products

Lucía Molina-García, Eulogio José Llorent-Martínez and María Luisa Fernández-de Córdova*
Department of Physical and Analytical Chemistry,
Faculty of Experimental Sciences, University of Jaén, Jaén, Spain

Abstract

Cocoa (*Theobroma cacao*), from the health viewpoint, is an important commodity in the world due to the high contents of polyphenols which are its main antioxidant-active fraction, particularly flavonoids. Polyphenols also confer astringent and bitter taste, also contributing to the fruity and green flavors of cocoa liquors. There are three main groups of flavonoids in cocoa: catechins (flavan-3-ols), anthocyanins, and proanthocyanidins. Other minor polyphenols also present in cocoa include flavones, phenolic acids, clovamide, dideoxyclovamide, resveratrol, and piceid. The content and composition

---

* Corresponding Author Email: mferna@ujaen.es.

of polyphenols in cocoa and its derived products vary strongly depending on genotype, cultivar, origin, agricultural practices, post-harvest practices, and bean processing. Fermentation of cocoa beans is considered to be one of the major steps that affect polyphenol content. Anthocyanins usually disappear rapidly during this process being hydrolyzed to anthocyanidins, which polymerize along with simple catechins to form complex tannins. During the drying process, the oxidation reactions that started in fermentation continue, the phenol degradation increasing with temperature. Alkalinization of the nibs, liquor, or powder also originates considerable losses of cocoa polyphenols. Therefore, given the functional effects of the polyphenols above mentioned, it is vital analyzing the total phenolic content and the composition and quantity of individual polyphenols in both cocoa beans and cocoa-derived products, being chocolate the most representative of these latter.

The quantitative determination of total polyphenols in cocoa and chocolate usually includes the assessment of the total phenol (TPC) or flavonoid (TFC) contents with *Folin-Ciocalteu* reagent, the measurement of the flavan-3-ols content by the vanillin and *p*-dimethylaminocinnamaldehyde (*p*-DAC) assays or the determination of proanthocyanidins by the butanol/HCl assay. The analysis of the polyphenols profile has usually been accomplished by liquid chromatography (HPLC) with different detectors. Nevertheless, nowadays liquid chromatography-mass spectrometry (HPLC-MS) or liquid chromatography-tandem mass spectrometry (HPLC-MS/MS) are among the most widely used techniques for both structural characterization and quantification of both low and high molecular weight polyphenols in cocoa. In the last years, there have been significant improvements in HPLC technique with the development of ultra-performance liquid chromatography (UPLC) and today, the number of studies focusing on the analysis of polyphenols by using UPLC coupled to MS (UPLC-MS/MS) is increasing. In this chapter, the different current available methodologies for analysis, quantification, isolation, purification, and structure elucidation of polyphenols in cocoa and cocoa-derived products are reviewed. Taking into account that there is evidence that demonstrates that the bioactivity of flavanols is significantly influenced by their stereochemical configuration, enantioselective methods have also been included such as chiral capillary electrophoresis (CCE), micellar electrokinetic chromatography (MEKC) or UPLC.

Keywords: cocoa, chocolate, cocoa products, polyphenols, analytical methods

INTRODUCTION

Polyphenols in cocoa (*Theobroma cacao* L.) and cocoa products are mainly flavonoids, compounds consisting of two phenyl rings linked by three carbon atoms, a structure usually referred to as C6-C3-C6. Flavonoids are especially important antioxidants due to their high redox potential, which allows them to act as reducing agents, hydrogen donors, and singlet oxygen quenchers (Tsao and Yang, 2003). Cocoa flavonoids can be classified into three main groups: flavan-3-ols or catechins (~37%), proanthocyanidins (~58%) and anthocyanins (~4%) (Wollgast and Anklam, 2000).

Catechins are monomeric forms of flavanols, derivatives of flavans, and in cocoa products are mainly represented by (–)-epicatechin (up to 35% of the total polyphenols), (+)-catechin, (+)-gallocatechin, and (–)-epigallocatechin. Catechins can also exist as condensed polymers, the so-called proanthocyanidins. These flavonoids are also known as condensed tannins and are dimers, oligomers and polymers of catechins that are bound together by links between C4 and C8 (or C6) (Lucci et al., 2017). The most important proanthocyanidins found in cocoa beans are B1 (epicatechin-(4β→8)-catechin), B2 (epicatechin-(4β→8)-epicatechin), B3 (catechin-(4α→8)-catechin), B4 (catechin-(4α→8)-epicatechin), B5 (epicatechin-(4β→6)-epicatechin), C1 (epicatechin-(4β→8)-epicatechin-(4β→8)-epicatechin), and D (epicatechin-(4β→8)-epicatechin-(4β→8)-epicatechin-(4β→8)-epicatechin) (Aprotosoaie et al., 2016). Oligomeric proanthocyanidins are soluble in different aqueous and organic solvents; however, high-molecular weight condensed tannins are insoluble. In addition, condensed tannins can form insoluble complexes with protein or cell wall polysaccharides. These characteristics of proanthocyanidins have to be taken into account in the extraction of polyphenols from cocoa products because they can be an important source of error (Bravo, 1998). Anthocyanins are glycosides of anthocyanidins and the most important group of water-soluble plant pigments. The anthocyanin fraction of cocoa beans is constituted by leucoanthocyanins L1, L2, L3, and L4, cyanidin-3-

α-L-arabinoside, and cyanidin-3-β-D-galactoside (Aprotosoaie et al., 2016).

Sanbongi et al., also reported the occurrence of quercetin, (–)N-[3'-4'-dihydroxy-(E)-cinnamoyl]-(L)-dihydroxyphenylalanine (clovamide), (–)-N-[4'-hydroxy-(E)-cinnamoyl]-L-tyrosine (deoxyclovamide), quercetin 3-*O*-β-D-glucopyranoside, and quercetin 3-*O*-β-D-arabinopyranoside in cocoa liquor, which is the liquid (or paste) produced during chocolate manufacturing, after cocoa beans are roasted and finely ground. Other minor polyphenols also found in cocoa samples are hyperoside (quercetin-3-*O*-galactoside), luteolin, naringenin, apigenin and some *O*-glucosides and *C*-glucosides of these compounds, such as luteolin-7-*O*-glucoside, naringenin-7-*O*-glucoside, isoorientin (luteolin-6-*C*-glucoside), orientin (luteolin-8-*C*-glucoside), vitexin (apigenin-8-*C*-glucoside) and isovitexin (apigenin-6-*C*-glucoside), which were identified for the first time by Sánchez-Rabaneda et al.

In 2006, the stilbenes *trans*-resveratrol and *trans*-piceid were identified for the first time in dark chocolate and cocoa liquor extracts (Counet et al., 2006). These compounds are characterized by a double bond connecting two aromatic rings in a C6-C2-C6 structure with several hydroxyl groups. Therefore, both *cis* and *trans* isomers naturally occur, being the *trans* isomer comparatively more common. Both resveratrol and its glycoside, piceid, are relevant compounds because of their antioxidant properties. Nevertheless, the authors reported that the exceptional antioxidant activity of chocolate must be related more to its high proanthocyanidin content than to the presence of stilbenes.

Factors affecting the quality and quantity of cocoa and cocoa-derived products during their production and manufacturing are of great importance in delivering the best health effects. These factors could significantly reduce the polyphenol content of the selected products. Genetic variability, ecological conditions in the cocoa-growing regions, and the variety of microclimates lead to cocoa production with high variability in its chemical composition and, particularly, to differences in the polyphenol content of fresh cocoa beans (Clapperton et al., 1994; Carrillo et al., 2014). It is a well-known fact that the processing of cocoa beans also significantly influences the polyphenol content of cocoa products. After harvesting, cocoa beans undergo complex processes such as fermentation and drying that alter their original chemical and physical properties in order to increase the palatability of seeds and obtain chocolate flavors. During the cocoa beans fermentation process, a series of biochemical reactions occur that cause a reduction of the astringency and bitterness and originate the precursors of the flavor of chocolate. In the process, acids such as acetic and lactic acid are generated, which increase acidity and cause hydrolysis reactions. Concurrently an oxidation of polyphenols to *o*-quinones by the enzyme polyphenol oxidase takes place. The oxidative process started during fermentation continues during the drying step, particularly if this latter is performed at temperatures above 60ºC (Kyi et al., 2005) and the *o*-quinones generated subsequently react with anthocyanins to form brown polymers. The secondary processing involves roasting and alkalization. Roasting is the principal technological operation in cocoa beans processing that affects the composition of the roasted seeds, and defines the quality and sensory characteristics of the final cocoa-derived products. During this step of cocoa beans processing, the high temperatures used (130ºC-150°C) also induce significant changes in the composition of polyphenolic compounds, which are readily degraded and/or become bound to polymer structures (Oracz et al., 2015). Alkalization, also known as "Dutching," consists in the treatment of cocoa mass, liquor, or powder with alkali and it is carried out to improve the color and flavor of cocoa (Aprotosoaie et al., 2016). Miller et al., studied the loss of polyphenols during this treatment and concluded that about 40% of the natural level of flavanols is retained on average for lightly Dutched powders. In the case of moderately alkalized cocoa powders, the retention percentage is still about 22%.

Polyphenols constitute a family of bioactive compounds with a great variety of positive health effects, including anti-inflammatory, anti-carcinogenic, anti-thrombotic, anti-microbial, anti-ulcer and vasodilatory effects that have caught the attention of consumers over the last few years (Jonfia-Essien et al., 2008). Therefore, their isolation, purification, and qualitative and quantitative analysis in cocoa-derived products are of relevance for society, food producers, and researchers. The analysis of polyphenols in cocoa samples is relatively complicated due to the great variety of compounds that can be present, but also because many of these compounds are present at low concentration levels. All this has hindered their sample extraction and treatment as well as their separation, determination, and identification. Regarding the sample treatment, solid-liquid extraction and solid-phase extraction (SPE) are currently among the most used techniques for the fractionation/purification step in analysis of cocoa polyphenols, although other extraction approaches have also been tested. On the other hand, liquid chromatography coupled with mass spectrometry (HPLC-MS) or tandem mass spectrometry (HPLC-MS/MS) is the most effective technique for the structural characterization and determination of both high and low molecular weight polyphenols in food samples (Motilva et al., 2013; Wollgast and Anklam, 2000; Lucci et al., 2017). A significant limitation of these techniques to take into account when high throughput analysis is required is the longtime of analysis. Recently, ultra-high performance liquid chromatography (UPLC) has offered new possibilities for improving the analytical methodologies for complex food matrices, providing faster separations than conventional HPLC technique while maintaining or even increasing resolution (Lucci et al., 2017).

This chapter aims at carrying out a revision and assessment of the main analytical methodologies proposed for qualitative and/or quantitative analysis of polyphenols in cocoa and cocoa-derived products, as well as for their structure elucidation and characterization. Since the enantiomers of flavan-3-ols present different physiological activities, enantioselective methodologies reported for this purpose will also be reviewed, such as UPLC, chiral capillary electrophoresis (CCE) and micellar electrokinetic chromatography (MEKC).

SAMPLE PREPARATION

Since some cocoa phenolics are unstable, as is the case of anthocyanins and proanthocyanidins, care is required during the storage of the sample. For quantification purposes, snap freezing in liquid nitrogen is very advisable (Markham and Bloor, 1998). The first step of the analysis is the grinding and homogenization of the sample for ensuring an appropriate contact with the extraction solvent. An acid or alkaline hydrolysis can also be required to free some bound phenolics from their esters or glycoside forms, after or before the extraction step.

Before the extraction of phenolics, a complete removal of lipids (defatting) is usually necessary to accomplish maximum-efficiency extraction, primarily of procyanidins (Adamson et al., 1999). Lipid elimination is usually carried out by extraction with *n*-hexane (Hammerstone et al., 1999; Pedan et al., 2016), although extraction with *n*-heptane (Patras et al., 2014) and Soxhlet extraction with diethyl ether have also been proposed (Counet and Collin, 2003). The non-polar solvent allows to remove co-extracted lipophilic compounds efficiently that would otherwise cause a subsequent reduction in ionization efficiency and/or chromatographic separation problems. Automatic Soxhlet extraction has also been proposed for both to remove the lipid fraction (12 h), using dichloromethane, and to extract polyphenols with methanol as solvent (4 h) from cocoa beans, roasted nibs and hulls (Arlorio et al., 2008).

The selection of an appropriate extraction solvent is a determining step in the analysis of polyphenols. The extraction procedure obviously will depend on the nature of the sample matrix as well as on the chemical properties of the phenolics. In HPLC-MS analysis, the solvent will also determine the detection limits (LOD) and matrix effects. Although some authors have used water and ethanol as extractants, oligomeric procyanidins are extracted only partially, and high molecular weight polymers are not extracted at all (Wollgast and Anklam, 2000). The most commonly used solvents for the extraction of cocoa polyphenols are 70-80% aqueous methanol (Sánchez Rabaneda et al., 2003) and 70-80% aqueous acetone (Natsume et al., 2000) or combinations thereof, since they are generally more soluble in polar organic solvents than in pure water. Furthermore, compounds other than phenolics such as water-soluble proteins, peptides, carbohydrates, and organic acids may be co-extracted when increasing the water amount in the extraction solvent or when using water alone (Boeing et al., 2014). Acetone-based solvents are more efficient for extracting oligomeric procyanidins than other extraction solvents, e.g., alcohol/water mixtures.

In the extraction of flavanols, it is necessary to take into account that a condensation of low molecular weight polyphenols can occur to form tannins of a higher polymerization degree. In addition, during fermentation of cacao beans, polyphenols can form complexes with proteins and carbohydrates and thus make it difficult to achieve efficient extraction from the matrix (Wollgast and Anklam, 2000). Consequently, the extraction of high molecular weight procyanidins could require the use of extraction methods with acid-catalyzed hydrolysis or higher temperatures, although labile monomers could be degraded in these aggressive working conditions. The use of acetone as extraction solvent may improve the extraction yield of polyphenols by inhibiting the formation of protein-polyphenol complexes during extraction or even by breaking down hydrogen bonds formed between phenolic groups and protein carboxyl groups (Chirinos et al., 2007).

Hammerstone et al., proposed the extraction of procyanidins from cocoa beans with 70% aqueous acetone followed by extraction with 70% aqueous methanol, while procyanidins of chocolate were extracted with acetone:water:acetic acid mixture (70:29.5:0.5, v/v/v). Counet and Collin optimized this solvent mixture and proposed a slightly modified version containing 2% acetic acid instead of 0.5%, which provided the best recovery of P1-P6 oligomeric procyanidins. A range of solvents, including methanol, ethanol, 1-propanol, 2-propanol, and acetone, and their respective mixtures with water were tested by Pedan et al., who found that a 50% aqueous acetone mixture showed the best efficiency for extracting the different polyphenol classes in cocoa.

Stilbenes are highly sensitive to light and, therefore, a specific extraction procedure is required to recover them from cocoa samples. Callemien et al., developed a procedure for the extraction of polyphenols from hop pellets, more adapted to the extraction of stilbenes (darkness, delipidation with a non-ether solvent), that was applied to their determination in cocoa products by Counet et al., The method involved delipidation with toluene and cyclohexane and extraction with 80% ethanol at 60ºC, with protection against light during analysis.

The extraction of polyphenols with conventional methods involves the use of large amounts of organic solvents and a previous defatting of the sample. In addition, these methodologies are labor-intensive and time-consuming. Therefore, an important aspect to consider is the development of fast, cost-effective, and environmental-friendly extraction methods that can isolate the polyphenols of interest from cocoa and cocoa products. Pressurized liquid extraction (PLE) is a technique involving extraction using liquid solvent subjected to elevated temperature and pressure, which enhances the extraction performance as compared to those techniques carried out at near room temperature and atmospheric pressure (Camel, 2001). The application of this technique improves the extraction yield, protects sensitive compounds, and decreases time and solvent consumption. When liquid water is used as extractant at temperatures above the atmospheric boiling point of water, the technique is called pressurized hot water extraction (PHWE). Recently, Plaza et al., developed a PHWE method that was more efficient in extracting phenolic compounds from cocoa and cocoa products (cocoa powder; chocolate; nibs; nibs powder) as compared to a conventional extraction method (Wollgast et al., 2001; Brcanovic et al., 2013). In the PHWE method the extractions were performed on a Dionex ASE 200 system (Thermo Fisher, Germering, Germany) at 125 °C for 3 min, while the conventional extraction method consisted in a previous elimination of lips with *n*-hexane and the posterior extraction of polyphenols with a mixture of acetone:water:acetic acid (70:29.8:0.2, v/v/v) at 30 °C in an ultrasonic bath. Also, the defatting of samples and the use of organic solvents were avoided. The differences between the extraction efficiencies found in both methods were higher in the case of procyanidins for all the analyzed samples and significantly higher for cocoa powder and nibs (nibs and nib powder) samples.

The application of the supercritical fluid extraction (SFE) technique has also been tested by using supercritical CO_2 for the extraction of procyanidins from cocoa seeds (Sarmento et al., 2008). As reported by the authors, this solvent was not effective since it has nonpolar properties, while polyphenols are compounds with polar properties. The extraction efficiency could be improved up to 40% by adding ethanol as a co-solvent.

After the extraction of polyphenols, a concentration step can be required, which is almost always done at low temperatures (below 40 ºC) and under reduced pressure in order to minimize the component degradation. Clean-up procedures are generally carried out as a previous step to further analysis by HPLC or capillary electrophoresis (CE) for removing potential interfering species such as sugars, caffeine, and theobromine. Removal of theobromine and caffeine can usually be accomplished by extraction with chloroform or dichloromethane, since most flavonoids have very limited solubility in these solvents. Procedures for purification of the extracts include semi-preparative HPLC, column chromatography, gel permeation chromatography (GPC) and SPE. Sephadex LH-20 has been widely used as adsorbent for further purification of flavonoids in column chromatography (Porter et al., 1991; Romanczyk et al., 1997; Sánchez-Rabaneda et al., 2003) and GPC (Adamson et al., 1999; Pedan et al., 2016). This adsorbent is a highly cross-linked dextran gel on which separations are obtained by molecular size combined with adsorption and, in non-aqueous solvents, an efficient, high-capacity medium for analytical and preparative work.

SPE is one of the most used techniques for the fractionation and purification of polyphenol extracts, as it offers the additional advantages of being rapid, economical, and straightforward to use. Furthermore, a great variety of solid supports are currently available as sorbents, and the technique can be easily automated to improve throughput. With regard to SPE stationary phases, octadecyl bonded silica (C18) has been by far the most common choice for extraction of flavonoids from cocoa samples (Hammerstone et al., 1999; Counet and Collin, 2003; Ortega et al., 2008). Counet and Collin carried out the purification of procyanidin extracts by SPE using C18 solid-support, sugars being removed with water. Procyanidins were eluted with acetone, methanol or ethanol. For samples with high water content, mixtures (70:30, v/v) of these organic solvents with water were used for elution purposes. The purification of anthocyanidins can be also carried out using C18 sorbent and pure methanol as eluent (Niemenak et al., 2006). Polyamide has also been successfully used for isolation and purification of flavan-3-ols from cocoa beans, cocoa powder, cocoa liquor and chocolate by SPE (Kofink et al., 2007; Kothe et al., 2013). The monomeric flavan-3-ols were eluted from the SPE cartridges with water and methanol and no problems of comigration or coelution of other similar substances were observed in the subsequent analysis by CCE.

Spectrophotometric Methods for Quantification of Total Phenolics and Their Classes

Basically, there are two general approaches for the determination of the phenolic content in cocoa and chocolate. The first approach, discussed in this Section, includes the determination of total phenolics and their classes, such as flavan-3-ols, flavonoids, and proanthocyanidins. In this case, it is not necessary the separation of the individual phenolics, and the determination is usually carried out by spectrophotometry, this technique being the most used for quantification of different classes of polyphenols due to its simplicity and low cost. Several spectrophotometric methods have been developed for the quantification of plant phenolics, which are based on different principles and are used to determine different structural groups. Table 1 shows some of the values reported in literature for total phenolics and content in several classes of phenolics found in cocoa beans and cocoa products by using different assays, which will be discussed in this Section.

Table 1. Polyphenol content found in cocoa beans and cocoa products

Sample	Total polyphenol content (TPC)	Total flavonoid content (TFC)	Total proanthocyanidin content (TPAC)[c]	Total flavanol content (TF)[d]	Reference
Instant cocoa powder	2.86 ± 11.7 mg GAE g^{-1} [a,e]	-	-	-	Bonhevi and Coll, 1997
Cocoa powder	611 mg GAE g^{-1} [a]	564 mg ECE g^{-1} [b]	-	-	Lee et al., 2003
Cocoa beans	67 - 149.4 mg NF-ECE g^{-1} [a]	-	-	-	Niemenak et al., 2006
Cocoa powder (conventional) cocoa powder (enriched) cocoa beans	64.3 - 84.2 mg GAE g^{-1} [a]	-	22.99 ± 1.08 mg g^{-1} [e,g]	-	Tomás-Barberán et al., 2007
Cocoa beans cocoa liquor cocoa powder cocoa nibs	399.9 ± 28.1 mg L-CAE g^{-1} [a,e] 378.8 ± 11.3 mg L-CAE g^{-1} [a,e] 296.5 ± 22.2 mg L-CAE g^{-1} [a,e] 437.2 ± 25.1 mg L-CAE g^{-1} [a,e]	- - - -	- - - -	292.2 ± 7.2 mg L-CAE g^{-1} [e] 275.9 ± 3.9 mg L-CAE g^{-1} [e] 117.2 ± 7.4 mg L-CAE g^{-1} [e] 306.6 ± 3.9 mg L-CAE g^{-1} [e]	Ortega et al., 2008
Cocoa powder cocoa liquor chocolate	3.48 - 33.50 mg NF-GAE g^{-1} [a]	1.25 - 16.60 mg NF-GAE g^{-1} [a]	0.23 - 4.75 mg NF-CyE g^{-1}	0.63 - 10.66 NF-CAE g^{-1}	Belščak et al., 2009

Sample	Total polyphenol content (TFC)	Total flavonoid content (TFC)	Total proanthocyanidin content (TPAC)[c]	Total flavanol Content (TF)[d]	Reference
Dark chocolate (71% cocoa)	18.26 ± 0.03 mg CAE g^{-1} [e,f]	6.3 ± 0.7 mg CAE g^{-1} [e,f]	-	-	Araujo Pimentel et al., 2010
dark chocolate (40% cocoa)	14.5 ± 0.5 mg CAE g^{-1} [e,f]	5.0 ± 0.4 mg CAE g^{-1} [e,f]	-	-	
milk chocolate (34% cocoa)	11.4 ± 0.5 mg CAE g^{-1} [e,f]	2.4 ± 0.2 mg CAE g^{-1} [e,f]	-	-	
white chocolate (30% cocoa)	8.2 ± 0.8 mg CAE g^{-1} [e,f]	1.0 ± 0.4 mg CAE g^{-1} [e,f]	-	-	
Milk chocolate	2.70 ± 0.39 mg GAE g^{-1} [a,e]	1.6 ± 0.2 mg CAE g^{-1} [b,e]	0.69 ± 0.18 mg CyE g^{-1} [e]	-	Todorovic et al., 2015
dark chocolate	11.99 ± 2.11 mg GAE g^{-1} [a,e]	5.5 ± 1.5 mg CAE g^{-1} [b,e]	2.31 ± 0.80 mg CyE g^{-1} [e]	-	
dark chocolate (raspberry)	11.56 ± 0.54 mg GAE g^{-1} [a,e]	5.8 ± 0.1 mg CAE g^{-1} [b,e]	2.27 ± 0.21 mg GAE g^{-1} [e]	-	
cocoa powder	35.35 ± 2.11 mg GAE g^{-1} [a,e]	18.37 ± 1.51 mg CAE g^{-1} [b,e]	7.07 ± 0.44 mg CyE g^{-1} [e]	-	
Cocoa beans	60.4 mg NFD-ECE g^{-1} [a]	up to 37.2 mg NFD-ECE g^{-1} [b]	up to 25.4 mg NFD-CyE g^{-1}	-	Pedan et al., 2016

[a]: Folin-Ciocalteu assay; [b]: aluminum chloride assay; [c]: butanol/HCl assay; [d]: vanillin assay; [e]: average value ± standard deviation (SD); [f]: Vinson et al., 2001; [g]: HPLC-MS/MS; GAE: mg of gallic acid equivalents per g of product; ECE: mg of epicatechin equivalents per g of product; CAE: mg of catechin equivalents per g of product; NF-CAE: mg of (+)-catechin equivalents per g of non-fatty product; L-CAE: mg of catechin equivalents per g of lyophilized extract; CyE: mg of cyanidin equivalents per g of product; NF-GAE: mg of gallic acid equivalents per g of non-fatty product; NF-CyE: mg of cyanidin equivalents per g of non-fatty product; NFD-CyE: mg of cyanidin equivalents per g of non-fatty dry matter; NFD-ECE: mg of epicatechin equivalents per g of non-fatty dry matter.

The second approach, discussed in the following Section, consists in the determination of the individual phenolic compounds after their separation. It is worth mention that a proper extraction of the phenolics is always required, no matter the methodology used for their determination. As above described, phenolics are mostly extracted with aqueous methanol or acetone after the elimination of lipids from the samples with *n*-hexane (Guyot et al., 1998; Hammerstone et al., 1999; Wollgast and Anklam, 2000).

Total Phenol Content (TPC)

According to literature data, TPC in cocoa products varies depending on the cocoa bean variety, geographical origin, ripeness and post-harvest conditions, such as fermentation, drying, roasting, processing, and storage (Wollgast and Anklam, 2000). In addition, TPC is dependent on the procedure used for its determination, and especially on the solvent used for phenolics extraction. The spectrophotometric *Folin-Ciocalteu* assay is the most commonly used procedure for the estimation of the TPC in food extracts, and it has been successfully applied to cocoa and cocoa-derived products (Serra Bonvehí and Ventura Coll, 1997; Belščak et al., 2009; Carrillo et al., 2014). Although this method does not provide a quantitative measurement of individual compounds, it is widely used because of its simplicity and use of standard equipment, and it has produced a large body of comparable data (Singleton et al., 1999). This assay is based on electron transfer reactions between the *Folin-Ciocalteu* reagent and phenolics in Na_2CO_3 solution and, under the reaction conditions, all the compounds readily oxidizable are determined. Complex polymeric ions formed from phosphomolybdic and phosphotungstic heteropoly acids are reduced by phenolic compounds to the molybdenum-tungsten blue complex, which is determined spectrophotometrically at 750-765 nm against a blank sample after 2 h of storage in the dark. Other readily oxidizable substances that are not phenols can also react, such as reducing sugars, sulfites, organic acids or ascorbic acid, interfering in the determination of TPC (Sánchez-Rangel et al., 2013). Different approaches to eliminate the contribution of interfering species during the determination of TPC in plant food extracts by the *Folin-Ciocalteu* assay have been proposed. These include: (i) a previous purification of the phenolic extracts by SPE (Georgé et al., 2005); (ii) the calculation of a corrected TPC value by subtracting the ascorbic acid reducing activity from the TPC quantified in the plant food extract (Isabelle et al., 2010) and (iii) the treatment of the phenolic extracts with oxidizing reagents such as hydrogen peroxide (H_2O_2) at levels that oxidize the interfering compounds (Ford et al., 2010).

A common difficulty related to colorimetric methods of analysis as *Folin-Ciocalteu* assay is the selection of a suitable standard, which is a consequence of the heterogeneity of the phenolic fraction. Gallic acid (GA) is frequently used as standard, and the results are expressed in mg of gallic acid equivalents (GAE) per g of defatted product (Kramling and Singleton, 1969). Because the reaction is quantitative, independent, and predictable, the analysis of a mixture of phenols can be recalculated in terms of any other standard. Catechin (CA) and epicatechin (EC) have also been used as calibration standards and final results expressed as mg of catechin or epicatechin equivalents (CAE/ECE) per g of defatted matter (Ortega et al., 2008; Pedan et al., 2016). It is necessary to take into account that sometimes the TPC data reported by different authors are not comparable due to they are expressed as mg of standard compound per g of cocoa product, mg of standard compound per g of defatted cocoa product or mg of standard compound per g of lyophilized extract.

The use of the *Folin-Ciocalteu* reagent in an alkali medium takes more than 2 h to perform the two-step analysis. Vinson et al., developed a simpler and faster procedure using a stable and commercially available reagent that contained phosphoric acid; the assay was carried out under acidic conditions in which polyphenols are more stable. In these acidic conditions, sugars do not interfere at the concentrations found in high sugar foods such as chocolate, whereas they interfere in the alkaline *Folin-Ciocalteu* method if present at high concentrations. Araujo Pimentel et al., used this modified method for the assessment of TPC in different types of chocolate (white, milk, and dark), and, as expected, the highest value was found in dark chocolate sample (71% cocoa).

TPC and antioxidant capacity of cocoa were compared to those of green tea, black tea, and red wine, products that are widely consumed, known to be rich in phenolic phytochemicals, and in which most of phenolics are other than flavonoids (Lee et al., 2003). Cocoa contained much higher levels of total phenolics (611 mg of GAE) per serving than black tea (124 mg of GAE), green tea (165 mg of GAE), and red wine (340 mg of GAE); in addition, cocoa showed the highest antioxidant capacity, these results suggesting that it is more beneficial to health than tea and red wine in terms of antioxidant content.

Belščak et al., analyzed different cocoa products (cocoa liquor, cocoa powder, cocoa bar, chocolate, cooking chocolate, powdered chocolate, and milk chocolate) and reported TPC values ranging from 3.48 mg GAE g^{-1} for milk chocolate to 33.50 mg GAE g^{-1} for defatted cocoa liquor. The extraction of phenolics compounds was carried out with 70% (v/v) aqueous methanol. According to these authors, the low TPC found in milk chocolate could be attributed to the lack of sensitivity of the *Folin-Ciocalteu* method when other reducing substances that are not phenolics are present in the sample and to the complex matrix of milk-based products where strong catechin-protein interactions are well-known to occur and can directly interfere with the accurate catechin determination. Also, it is necessary to take into account that, as above established, the phenolic content in cocoa products varies greatly depending on a lot of other factors. Serra Bonvehí and Ventura Coll estimated TPC values in cocoa powder of 58 mg GAE g^{-1} of product using the same extraction solvent. In the analysis of similar products, Waterhouse et al., reported TPC values of 20 mg GAE g^{-1} of cocoa powder, 8.4 mg GAE g^{-1} of dark chocolate, and 5 mg GAE $g^{-1}$of milk chocolate, using 95% (v/v) aqueous methanol as extraction solvent. Todorovic et al., used acetone:water:acetic acid (79:29.8:0.2, v/v/v) for the extraction of phenolics in dark chocolate, milk chocolate, and cocoa powder and they reported that the TPC was the highest in cocoa powder samples (35.35 mg GAE g^{-1}). Dark chocolate samples with or without added raspberries were similar regarding TPC value, with overall ranges from 7.21 to 12.65 mg GAE g^{-1}. TPC was 3 and 13 times lower in dark and milk chocolate than in cocoa powder, what can be attributed to the alkalization process of cocoa powder and the influence of added lipid and protein-rich ingredients of chocolate.

Total Flavonoid Content (TFC)

The most commonly applied method for determination of total flavonoid content in cocoa products is based on the complexation of these compounds with Al_3Cl (Lee et al., 2003; Todorovic et al., 2015; Pedan et al., 2016). After addition of Al (III) in the presence of $NaNO_2$ in alkaline medium, a yellow solution of complex is formed, which then turns immediately to red, and absorbance is measured at 510 nm (Pękal and Pyrzynska, 2014). A nitration of any aromatic ring bearing a catechol group with its three or four positions unsubstituted or not sterically blocked occurs during the reaction. In the presence of $NaNO_2$, Al_3Cl seems to be specific for rutin, luteolin and catechins, but also phenolic acids exhibit considerable absorbance. Catechin is preferred as standard compound in this procedure.

On the other hand, the determination of TFC can be also carried out by the indirect method using formaldehyde. This latter reacts with C-6 or C-8 on 5,7-dihydroxyflavonoids to form methyl derivatives that further react with other flavonoids also at positions C-6 and C-8 and generate insoluble compounds (Ough and Amerine, 1988). The condensed products generated in these reactions are removed by filtration, and the remaining phenolic compounds are determined by the above described *Folin-Ciocalteu* method or the modification of this latter proposed by Vinson et al. TFC can be determined from the difference between TPC and non-flavonoid content.

Total Flavanol Content (TF)

Vanillin Assay

In contrast to *Folin-Ciocalteu* assay, the vanillin assay is specific for flavan-3-ols, dihydrochalcones, and proanthocyanidins containing free meta-oriented hydroxy groups in the A-ring and a single bond at the 2,3-position of the C-ring (Butler et al., 1982). Although the vanillin reagent is not completely specific for flavanoids and a significant color development can be expected by dihydrochalcones, this is not a problem in cocoa samples because they do not contain these phenolic compounds (Table 1). The method is based on the condensation in an acidic solution of the nucleophilic sites of phenolics with protonated vanillin, a substituted benzaldehyde (4-hydroxy-3-methoxybenzaldehyde), to yield red complexes with maximum absorption at 500 nm (Burns, 1971) (Figure 1a). (+)-Catechin and sulfuric acid are usually used as reference standard and catalyst, respectively. The results are influenced by the type of solvent used, the nature and concentration of the acid, the reaction time, the vanillin concentration, the type of reference standard used, and temperature.

p-Dimethylaminocinnamaldehyde Assay

This method consists in the use of the chromogen *p*-dimethylaminocinnamaldehyde (DMACA), which specifically reacts with proanthocyanidins polymers, small oligomers, and the immediate proanthocyanidin precursors, flavan-3,4-diols and flavan-3-ols (Delcour and Janssens de Varebeke, 1985). The reaction of these flavanoids with DMACA originates the development of a green color (λ_{max}: 640 nm). The structural requirements for reaction with vanillin seem to be also applicable for reactivity with DMACA. Thus, it seems that DMACA reagent, like vanillin, only reacts with flavanoids containing meta-oriented di- or trihydroxy substituted A-rings and a single bond at the 2,3 position of the C-ring (Figure 2). (+)-Catechin is used as reference for quantification purposes.

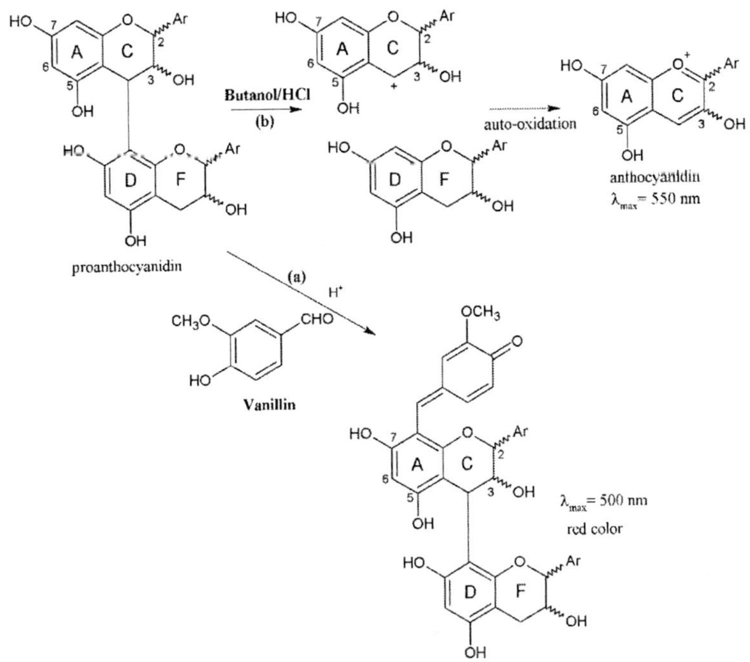

Figure 1. Chemical reactions involved in: (a) Vanillin assay for flavanols; (b) Butanol/HCl assay for proanthocyanidins.

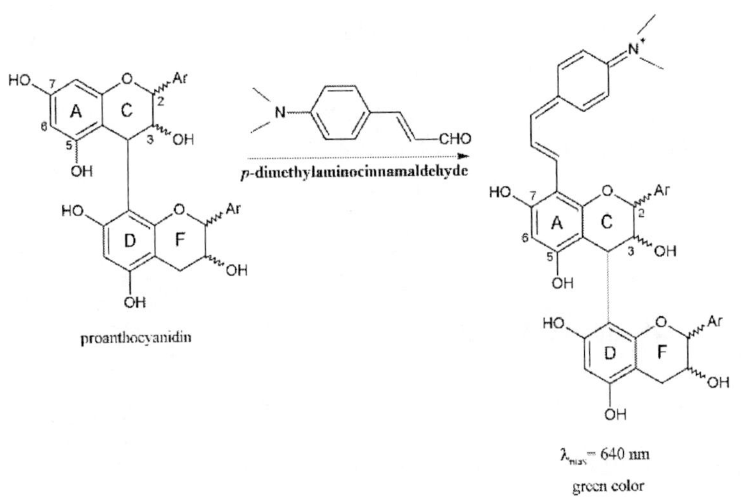

Figure 2. Chemical reaction involved in *p*-dimethylaminocinnamaldehyde assay for flavanols.

Total Proanthocyanidin Content (TPAC)

Butanol-HCl Assay

This is one of the most common colorimetric assays for determining the content in proanthocyanidins, flavonoids that, as is well known, cannot be extracted exhaustively from plant material. The assay is based on the oxidative depolymerization of proanthocyanidins in a hot acid-alcohol medium (Porter et al., 1986) and has been widely applied to cocoa samples (Table 1). First, proanthocyanidins suffer a depolymerization, being cleaved into the flavan-3-ol unit (terminal unit) and the carbocation unit (extension unit). The resulting anthocyanidins are mostly either cyanidin or delphinidin. Then, the carbocations are rapidly converted into red anthocyanidins by autoxidation. The assay is performed by mixing the sample with butanol-HCl and $NH_4Fe\ (SO_4)_2.12\ H_2O$ solutions, maintaining the mixture at 95°C for 40 min and measuring the absorbance at 550 nm (Figure 1b). The yield of anthocyanidin is affected by small amounts of water in the reaction mixture and by the presence of trace impurities of transition metal ions, which can be avoided by adding Fe^{3+} in order to favor the autoxidation step. Although this assay is simple and gives a good indication of the content in condensed tannins, the structure of these latter, and particularly the position of the interflavan bond, influences the yield and the kinetics of the reaction. In addition, the results of quantification strongly depend upon the standard used for calibration.

Analytical Methodologies for Analysis of Individual Phenolic Compounds

Among the analytical methodologies developed for the separation, quantitative determination and identification of polyphenols in cocoa and cocoa products, HPLC is the most applied technique due to its high resolution, high efficiency, high reproducibility, relatively short time of analysis without derivatization, and no restriction on sample volatility

(Wollgast and Anklam, 2000). Both reversed-phase HPLC (RP-HPLC) (Cooper et al., 2007; Andrés-Lacueva et al., 2008; Alañón et al., 2016) and normal-phase HPLC (NP-HPLC) (Adamson et al., 1999; Kealey et al., 1998; Counet and Collin, 2003) have been widely used for the analysis of flavan-3-ols and procyanidins. Due to the structural similarity of most polyphenols and the complex nature of cocoa matrix, most of the solvent systems used in HPLC methods include binary gradient elution and, occasionally, isocratic elution. Several detection systems have been explored for characterizing procyanidins in cocoa samples by HPLC; among them UV-Vis/diode-array (DAD) (Rigaud et al., 1993; Kealey et al., 1998) and fluorescence (Adamson et al., 1999; Cooper et al., 2007) have demonstrated to be very efficient.

The coupling of liquid chromatography to mass spectrometry (HPLC-MS) with atmospheric pressure ionization techniques, i.e., electrospray ionization (ESI) or atmospheric pressure chemical ionization (APCI), has demonstrated to be a powerful tool for structural elucidation due to its soft ionization, which favors the analysis of flavonoids, a polar, non-volatile, and thermally labile class of compounds (Sánchez Rabaneda et al., 2003). For instance, procyanidins were analyzed in cocoa and chocolate by RP-HPLC-ESI-MS (Wollgast et al., 2001; Natsume et al., 2000) and by NP-HPLC-ESI-MS (Hammerstone et al., 1999; Counet and Collin, 2003); monomeric flavanols and flavonols were determined by Andrés-Lacueva et al., by a RP-HPLC-APCI-MS method.

Recently, a series of fourteen novel phenolics including sulfated flavanols, A-type dimeric and trimeric proanthocyanidins, and flavanol-flavanone adducts were identified in cocoa bean extracts (Patras et al., 2014) using HPLC-ESI-MS analysis including both high resolution (TOF) and tandem MS (ion trap) analysis. These techniques proved to be efficient methods for the structural elucidation of these compounds.

In the last years, there has been an effective improvement in HPLC technique with the development of UPLC. This technique takes full advantage of chromatographic principles to perform separations using columns packed with smaller particles (1.7 μm) and/or at higher flow rates. This leads to a shorter time of analysis, higher peak efficiency, and higher resolution. RP-UPLC-MS/MS technique has been applied by Cooper et al., to the determination of six of the major polyphenols of chocolate: catechin, epicatechin, dimer B2 (epicatechin-4β-8-epicatechin), dimer B5 (epicatechin-4β-6-epicatechin), trimer C1 (epicatechin-4β-8-epicatechin-4β-8-epicatechin), and tetramer D (epicatechin-4β-8-epicatechin-4β-8-epicatechin-4β-8-epicatechin); the chromatographic analysis was carried out in 3 min, while the same separation by RP-HPLC takes over 20 min. Ortega et al., 2010 developed a sensitive UPLC-MS/MS method to identify and quantify flavan-3-ols and procyanidins in cocoa nibs, which was compared to HPLC-MS/MS method regarding sensitivity, selectivity, time of analysis, peak efficiency cost, and quality parameters. The UPLC-MS/MS method was developed by the authors by using RP or "aqueous normal" phase, since silica columns for UPLC in NP had not been developed yet, and it allowed the determination of procyanidins up to nonamers at low-concentration levels in less than 12.5 min, against 80 min for the HPLC-MS/MS method. Apart from the time of analysis, other differences were observed such as the resolution of the monomers catechin and epicatechin, that in the RP-UPLC methodology were baseline-separated but, on the contrary, they were not entirely separated in the NP-HPLC methodology.

Apart from HPLC, other techniques such as thin layer chromatography (TLC) and MEKC (Gotti et al., 2006) have also been successfully used for the analysis of polyphenols in cocoa samples. TLC is a technique with large applicability in the field of plant material analysis and stability tests of extracts and final products, and it shows advantages such as lower costs, short analysis time and the possibility of multiple detections. This technique has been used especially for qualitative analysis, but also for the screening of flavonoid extracts or fractions from other chromatographic separations (Porter et al., 1991).

Capillary electrophoresis (CE) is an alternative separation technique to HPLC and especially suitable for the separation and quantification of low to medium molecular weight polar and charged compounds, the resultant separations being often faster and more efficient than the corresponding HPLC separations (Caridi et al., 2007). This technique is increasingly becoming a versatile analytical tool for the routine determination of a wide variety of phenolics in different types of samples because of its high separation efficiency, high-resolution power, low consumption of sample and reagents, and high throughput (Ignat et al., 2011). Nevertheless, currently, HPLC remains the method of choice for flavonoid analysis. Probably, this is due to one of the major limitations of CE when comparing to HPLC: the collected UV spectra are often noisy due to the small amount of sample injected and LODs for compounds in the sample are higher. Since mobility in CE is determined mainly by the charge-to-mass ratio of the compound, flavonoids not carrying a charge must be ionized by means of an appropriate buffer, borate buffer (pH 8-11) being one of the most used as background electrolyte. However, owing to the poor stability of catechin monomers in alkaline solutions, other authors have preferred the use of Britton–Robinson buffer at pH 2.50 as the background electrolyte (Gotti et al., 2006). Flavonoids that are hydrophobic or are not readily ionized may be effectively analyzed by MEKC. In this technique a surfactant, often sodium-dodecyl sulfate (SDS), is introduced to generate charged micelles in which the neutral flavonoids can migrate (Wollgast and Anklam, 2000). MEKC has been successfully applied to the analysis of catechins from cocoa (Romanczyk et al., 1997; Gotti et al., 2004); the major advantage of these applications concerns the short analysis time in comparison with those of the proposed HPLC systems, which often required gradient elution. CE and MEKC have also been applied to the chiral separation of catechins enantiomers; these applications will be discussed in the next Section.

Recently, the ability of FT-IR technique combined with multivariate statistical analysis for quantifying and predicting TPC and (+)-catechin and (−)-epicatechin contents in chocolate has been tested (Hu et al., 2016). Partial least-squares regression (PLSR) models were constructed and cross-validated by correlating FT-IR spectral features from 18 types of chocolate with the corresponding reference values measured using chemical-based assays, *Folin-Ciocalteu* and HPLC methods. Then, the models were validated by using seven unknown samples of chocolate. The results obtained showed that both phenolic and (+)-catechin content of chocolate could be predicted from FT-IR spectral features. Nevertheless, PLSR models lacked accuracy in the prediction of (–)-epicatechin. The authors attributed the differences in the model performance for (+)-catechin and (–)-epicatechin to different losses for both catechins during the processing of chocolate. The analysis could be performed in 2 min and without any sample pre-treatment. Therefore, the combination of FT-IR with chemometrics provides new opportunities for the development of more robust and precise prediction models and methods able to be applied for determining chocolate content. Some separation methodologies for determining phenolics from cocoa beans and cocoa products reported in the last years are presented in Table 2.

ASSESSMENT OF FLAVANOL STEREOISOMERS

As above described, cocoa is a rich source of flavonoids, particularly flavan-3-ols, predominating catechin and epicatechin, which possess two chiral centers. As is well known, bioactivity of organic compounds depends significantly on the conformation of their molecules and, therefore, the distinction of stereochemical forms of flavan-3-ols is relevant. Among stereoisomers (+)-catechin and (–)-epicatechin are most likely be regarded as naturally occurring forms of cocoa flavanols, this latter having been identified as the more active compound responsible for the vascular health benefits associated with the consumption of cocoa and chocolate (Schroeter et al., 2006). The atypical flavan-3-ol enantiomers (–)-catechin and (+)-epicatechin rarely occur in nature and their presence seems to be induced by enzymatic action (Ellis et al., 1983). (+)-Catechin and (–)-epicatechin are found present in unfermented, dried and unroasted cocoa beans; nevertheless, during the manufacturing process of cocoa beans epimerization reactions can occur as a consequence of high temperatures during the roasting process, and particularly of the alkalization treatment of cocoa powder. The reaction mechanism is not fully clarified, but it is assumed that ring opening occurs on position C-2 of the oxygenated ring, and reclosing leads to the atypical enantiomers (Kothe et al., 2013). Consequently, in roasted cocoa beans and cocoa products the atypical stereoisomers (−)-catechin and (+)-epicatechin can also be found, due to the conversion of (−)-epicatechin in its epimer (−)-catechin, and similarly that of (+)-catechin in (+)-epicatechin (Kofink et al., 2007).

In literature, numerous analytical methods have been developed for the characterization and quantification of polyphenols from cocoa and cocoa products. However, only a few enantioselective analytical methodologies, including HPLC (Donovan et al., 2006; Machonis et al., 2012; Alañón et al., 2016), UPLC (Cooper et al., 2007), CCE (Kofink et al., 2007; Kothe et al., 2013), and MEKC (Gotti et al., 2006) have been reported. Table 3 shows the content in catechins and their enantiomers in cocoa beans and cocoa products reported by different authors.

Gotti et al., 2006 developed an MEKC method that allowed the separation of the enantiomers of catechin monomers, (−)-epicatechin, (+)-epicatechin, (−)-catechin and (+)-catechin by means of the addition of hydroxypropyl-β-cyclodextrin (HP-β-CD) to the SDS micellar solution, resulting in a cyclodextrin-modified MEKC (CD-MEKC) methodology. The enantioselectivity in these systems appears to be due to the formation of an inclusion complex between a solute and the CD (Chankvetadze, 1997). This method presented two peculiar advantages compared to the conventional MEKC: (i) strong improvement of separation of the most important phytomarkers of *Theobroma cacao* and (ii) enantioselectivity toward (±)-catechin. In this work, the native enantiomer (+)-catechin was found in cocoa beans and the authors reported for the first time the predominant presence of (−)-catechin in chocolate, probably yielded by epimerization of (−)-epicatechin occurred during the manufacture of this product.

Table 2. Analytical methodologies for isolation/purification, identification and quantification of polyphenols from cocoa beans and cocoa products

Sample	Extraction	Clean-up	Analytical methodology	Reference
Cocoa beans	60% aqueous acetone	Extraction with ethyl acetate; aqueous phase: column chromatography on Sephadex LH-20 or column chromatography on TSK-HW-40(F); ethyl acetate phase: column chromatography on Sephadex LH-20	TLC and 2-D TLC; Vanillin-HCl detection RP-HPLC; UV-Vis detection at 280 nm	Porter et al., 1991
Cocoa beans	70% aqueous methanol	Extraction with ethyl acetate	NP-HPLC; UV-Vis detection at 280 nm	Rigaud et al., 1993
Cocoa beans; chocolate; chocolate raw products	. 70% aqueous acetone . 70% aqueous methanol	Column chromatography on Sephadex LH-20	NP-HPLC; DAD detection at 280 nm; fluorescence detection (λ_{ex} 276 nm, λ_{em} 316 nm) RP-HPLC; DAD detection at 280 nm; fluorescence detection (λ_{ex} 276 nm, λ_{em} 316 nm) MEKC; DAD detection at 280 nm; fluorescence detection (λ_{ex} 276 nm, λ_{em} 316 nm)	Romanczyk et al., 1997
Cocoa beans; cocoa liquor	70:29.5:0.5 (v/v/v) acetone: water: acetic acid	-	NP-HPLC; DAD detection at 280 nm; fluorescence detection (λ_{ex} 276 nm, λ_{em} 316 nm)	Kealey et al., 1998
Cocoa beans; chocolate	70% aqueous acetone + 70% aqueous methanol (beans); 70:29.5:0.5 (v/v/v) acetone: water: acetic acid (chocolate)	SPE on C18 Baker packing (beans) or Supelcosil Envi-18 column (chocolate); sugars removed with water; eluent: 70:29.5:0.5 (v/v/v) acetone: water: acetic acid	NP-HPLC-ESI-MS	Hammerstone et al., 1999

Sample	Extraction	Clean-up	Analytical methodology	Reference
Cocoa beans; chocolate	70% aqueous acetone	GPC on Sephadex LH-20 column + preparative normal-phase HPLC (Supelcosil LC-Si 100 Å column)	NP-HPLC; fluorescence detection (λ_{ex} 276 nm, λ_{em} 316 nm)	Adamson et al., 1999
Cocoa beans; cocoa liquor; chocolate	80% aqueous acetone	-	RP-HPLC (catechins); UV-Vis detection at 280 nm NP-HPLC (procyanidins); UV-Vis detection at 280 nm RP-HPLC-ESI-MS (procyanidins)	Natsume et al., 2000
Chocolate	70:29.8:0.2 (v/v/v) acetone: water: acetic acid	-	RP-HPLC-ESI-MS/MS	Wollgast et al., 2001
Cocoa powder	80% aqueous methanol	Column chromatography on Sephadex LH-20	RP-HPLC-ESI-MS/MS	Sánchez-Rabaneda et al., 2003
Dark chocolate	70:28:2 (v/v/v) acetone: water: acetic acid	SPE on C18 Sep-Pack cartridges; sugars removed with water; eluent: methanol, ethanol, water (100%) or 70% aqueous organic solvent (samples with high water content)	NP-HPLC; DAD detection at 280 nm NP-HPLC-ESI-MS	Counet and Collin, 2003
Cocoa beans	60% aqueous acetone	Extraction with ethyl acetate (polyphenols) SPE on C18 Sep-Pack cartridges; eluent: methanol (anthocyanidins)	RP-HPLC; DAD detection at 280 nm (polyphenols) RP-HPLC; DAD detection at 520 nm (anthocyanidins)	Niemenak et al., 2006
Cocoa beans	29% aqueous ethanol	-	CD-MEKC; DAD detection at 220 nm	Gotti et al., 2006
Cocoa beans; cocoa liquor; cocoa powder; chocolate	80% aqueous methanol	SPE on Chromabond polyamide cartridges; eluent: methanol + water + methanol	CCE; DAD detection at 280 nm	Kofink et al., 2007

Table 2. (Continued)

Sample	Extraction	Clean-up	Analytical methodology	Reference
Cocoa powder; chocolate milk drinks	99.9:0.1 (v/v/v) methanol: HCl	-	RP-HPLC; DAD detection at 280 nm RP-HPLC-ESI-MS/MS	Tomás-Barberán et al., 2007
Chocolate	70:28:2 (v/v/v) acetone: water: acetic acid	-	RP-UPLC; DAD detection at 220 nm RP-UPLC-ESI-MS/MS Chiral HPLC; fluorescence detection (λ_{ex} 280 nm, λ_{em} 310 nm)	Cooper et al., 2007
Cocoa beans; cocoa liquor; cocoa powder; cocoa nibs	70:29.5:0.5 (v/v/v) acetone: water: acetic acid	SPE on C18 cartridges; eluent: 70:29.5:0.5 (v/v/v) acetone: water: acetic acid	RP-UPLC-ESI-MS/MS	Ortega et al., 2008
Cocoa powder	80% aqueous methanol	-	RP-HPLC; DAD detection at 280 and 365 nm RP-HPLC-APCI-MS	Andrés-Lacueva et al., 2008
Cocoa powder cocoa liquor chocolate	water or 70% methanol	-	RP-HPLC; DAD detection at 278 nm	Belščak et al., 2009
Cocoa nibs	70:29.5:0.5 (v/v/v) acetone: water: acetic acid	-	RP-UPLC-ESI-MS/MS NP-HPLC-ESI-MS/MS	Ortega et al., 2010
Cocoa beans	50% aqueous acetone (ASE)	SPE on Chromabond polyamide cartridges; eluent: 99.9/0.1 (v/v) dimethyl sulfoxide/formic acid	CCE; DAD detection at 280 nm RP-UPLC-ESI-MS/MS	Kothe et al., 2013

Sample	Extraction	Clean-up	Analytical methodology	Reference
Cocoa beans	70% aqueous methanol	-	RP-HPLC-ESI-MS/MS RP-HPLC-ESI-TOF MS	Patras et al., 2014
Chocolate; cocoa powder	70:29.8:0.2 (v/v/v) acetone: water: acetic acid	-	RP-HPLC; DAD detection at 278 nm	Todorovic et al., 2015
Cocoa beans	50% aqueous acetone	SCPC (biphasic solvent: ethyl acetate/butanol/water 4:1:5 v/v/v) + GPC on Sephadex LH-20	RP-HPLC-ESI-MS	Pedan et al., 2016
Chocolate	70:28:2 (v/v/v) acetone: water: acetic acid	-	RP-HPLC; DAD detection at 220 nm Chiral HPLC; fluorescence detection (λ_{ex} 280 nm, λ_{em} 310 nm)	Alañón et al., 2016

TLC: thin layer chromatography; 2-D TLC: two dimensional thin layer chromatography; HPLC: high-performance liquid chromatography; RP: reversed phase; NP: normal phase; DAD: diode array detector; MS: mass spectrometry; ESI: electrospray ionization; APCI: atmospheric pressure chemical ionization; TOF: time of flight; MEKC: micellar electrokinetic chromatography; CD-MEKC: cyclodextrin-modified micellar electrokinetic chromatography; CCE: chiral capillary electrophoresis; SPE: solid-phase extraction; GPC: gel permeation chromatography.

Table 3. Content in catechins and their enantiomers in cocoa beans and cocoa products

Sample	Epicatechin content (mg g^{-1})	Catechin content (mg g^{-1})	Reference
Instant cocoa powder	0.180-0.320 (−)-epicatechin[a]	-	Bonhevi and Coll, 1997
Chocolate cocoa powder cocoa powder (alkalinized)	0.18-1.25 epicatechin 1.580-2.580 (−)-epicatechin 0.180-0.380 (−)-epicatechin	0.05-0.33 catechin[a,g] 0.610-0.900 (+)-catechin[a,g] 0.230-0.350 (+)-catechin[a,g]	Gu et al., 2006
Cocoa beans	4.10-15.30 (−)-epicatechin[f]	0.06-0.45 (+)-catechin[f]	Gotti et al., 2006
Cocoa powder (conventional) cocoa powder (enriched)	3.30 ± 0.29 epicatechin[b] 25.65 ± 0.73 epicatechin[b]	2.02 ± 0.04 catechin[b] 6.46 ± 0.55 catechin[b]	Tomás-Barberán et al., 2007
Chocolate	0.679 ± 0.413 (−)-epicatechin[c] 0.033 ± 0.022 (+)-epicatechin[c]	0.218 ± 0.126 (−)-catechin[c] 0.025 ± 0.015 (+)-catechin[c]	Cooper et al., 2007
Cocoa powder	0.116-0.730 (−)-epicatechin[b]	0.081-0.448 (+)-catechin[b]	Andrés-Lacueva et al., 2008
Cocoa beans	1.16-4.77 (−)-epicatechin[d]	0.02-0.09 (+)-catechin[d] 0.02-0.18 (−)-catechin[d]	Kothe et al., 2013
Chocolate	0.081-0.895 (−)-epicatechin[a] 0.006-0.046 (+)-epicatechin[a]	0.018-0.601 (−)-catechin[a] 0.004-0.058 (+)-catechin[a]	Alañón et al., 2016

Catechins contents in table were obtained by: [a]: high-performance liquid chromatography (HPLC); [b]: high-performance liquid chromatography-tandem mass spectrometry (HPLC-MS/MS); [c]: ultra-high performance liquid chromatography-tandem mass spectrometry (UPLC-MS/MS); [d]: chiral capillary electrophoresis (CCE); and [f]: cyclodextrin-modified micellar electrokinetic chromatography (CD-MEKC); [g]: high-performance liquid chromatography- mass spectrometry (HPLC-MS).

Later, Cooper et al., developed a chiral HPLC methodology based on the use of a Cyclobond I-2000 RSP column packed with β-CD for the separation and fluorimetric detection (λ_{ex}, 280 nm; λ_{em}, 310 nm) of catechin enantiomers in chocolate. Epicatechin and catechin in chocolate were predominantly in the (−) form, which accounted for 95 ± 1% and 89 ± 3% of epicatechin and catechin present in the samples, respectively. The mean concentrations of (−)-epicatechin and (−)-catechin in chocolate samples were 0.679 ± 0.413 and 0.218 ± 0.126 mg/g, respectively. According to the authors, since (−)-catechin was predominant in all the chocolate samples, it may be the main form of polyphenols that varies with manufacturing conditions. Since in fresh cocoa beans the levels of (−)-catechin form are undetectable, the source of (−)-catechin in chocolate could be epimerization of the (−)-epicatechin during processing, as reported in a previous publication (Gotti et al., 2006). Kofink et al., performed the enantioseparation of catechin and epicatechin in cocoa and cocoa products by CCE with CD as chiral selectors and concluded that the first processing step for the formation of (−)-catechin is the roasting procedure, in which (−)-epicatechin is converted into (−)-catechin under the influence of high temperature. Alañón et al., analyzed dark and milk chocolate samples by using a chromatographic method reported in bibliography (Cooper et al., 2007) and found the naturally occurring forms of cocoa flavanols, (−)-epicatechin and (+)-catechin, together with the occurrence of (+)-epicatechin and (−)-catechin. (−)-epicatechin was the predominant form (93%).

Proanthocyanidins are composed of catechin and epicatechin units and also contain chiral centers; therefore, they could undergo epimerization reactions. Kothe et al., studied if epimerization of procyanidin dimers B1, B2, and B5 occurs during the roasting of cocoa beans and found that the concentration of the three dimers decreased with higher roast levels, showing progressive losses of 42%, 68%, and 71%, respectively, when heating at 160°C for 30 min. In contrast, the concentrations of two unknown dimers (dimer1 and dimer2) increased steadily with higher temperatures, 541% and 211% in the same roasting, and it was assumed that these dimers were products of the epimerization reaction.

The authors also separated catechins by CD-CCE and confirmed the epimerization of (−)-epicatechin to (−)-catechin.

References

Adamson, G. E.; Lazarus, S. A.; Mitchell, A. E.; Prior, R. L.; Cao, G. H.; Jacobs, P. H.; Kremers, B. G.; Hammerstone, J. F.; Rucker, R. B.; Ritter, K. A.; Schmitz, H. H. HPLC method for the quantification of procyanidins in cocoa and chocolate samples and correlation to total antioxidant capacity. *J. Agric. Food Chem.* 1999, *47*, 4184-4188.

Alañón, M. E.; Castle, S. M.; Siswanto, P. J.; Cifuentes-Gómez, T.; Spencer, J. P. E. Assessment of flavanol stereoisomers and caffeine and theobromine content in commercial chocolates. *Food Chem.* 2016, *208*, 177-184.

Andrés-Lacueva, C.; Monagas, M.; Khan, N.; Izquierdo-Pulido, M.; Urpi-Sarda, M.; Permanyer, J.; Lamuela-Raventós, R. M. Flavanol and flavonol contents of cocoa powder products: influence of the manufacturing process. *J. Agr. Food Chem.* 2008, *56*, 3111-3117.

Aprotosoaie, A. C.; Vlad Luca, S.; Miron, A. Flavor chemistry of cocoa and cocoa products-an overview. *Compr. Rev. Food Sci. F.* 2016, *15*, 73-91.

Araujo Pimentel, F.; Nitzke, J. A.; Blauth Klipel, C.; Vogt de Jong, E. Chocolate and red wine - A comparison between flavonoids content. *Food Chem.* 2010, *120*, 109-112.

Arlorio, M.; Locatelli, M.; Travaglia, F.; Coïsson, J-D.; Del Grosso, E.; Minassi, A.; Appendino, G.; Martelli, A. Roasting impact on the contents of clovamide (*N*-caffeoyl-*L*-DOPA) and the antioxidant activity of cocoa beans (*Theobroma cacao* L.). *Food Chem.* 2008, *106*, 967-975.

Belščak, A.; Draženka K.; Horžić, D.; Ganić, K. K.; Karlović, D. Comparative study of commercially available cocoa products in terms of their bioactive composition. *Food Res. Int.* 2009, *42*, 707-716.

Boeing, J. S.; Barizao, E. O.; Costa e Silva, B.; Montanher, P. F.; Almeida, V. D. C.; Visentainer, J. V. Evaluation of solvent effect on the extraction of phenolic compounds and antioxidant capacities from the berries: application of principal component analysis. *Chem. Cent. J.* 2014, *8*, 48/1-48/9.

Bonvehi, J. S.; Coll, F. V. Evaluation of bitterness and astringency of polyphenolic compounds in cocoa powder. *Food Chem.* 1997, *60*, 365-370.

Bravo, L. Polyphenols: chemistry, dietary sources, metabolism, and nutritional significance. *Nutr. Rev.* 1998, *56*, 317-333.

Brcanovic, J. M.; Pavlovic, A. N.; Mitic, S. S.; Stajanocic, G. S.; Manojlovic, D. D.; Kalicanin, B. M.; Veljkovic, J. N. Cyclic voltammetric determination of antioxidant capacity of cocoa powder, dark chocolate and milk chocolate samples: correlation with spectrophotometric assays and individual phenolic compounds. *Food Technol. Biotech.* 2013, *51*, 460-470.

Burns, R. E. Methods for estimation of tannin in grain sorghum. *Agron. J.* 1971, *63*, 511-512.

Butler, L. G.; Price, M. L.; Brotherton, J. E. Vanillin assay for proanthocyanidins (condensed tannins): modification of the solvent for estimation of the degree of polymerization. *J. Agr. Food Chem.* 1982, *30*, 1087-1089.

Callemien, D.; Jerkovic, V.; Rozenberg, R.; Collin, S. Hop as an interesting source of resveratrol for brewers: optimization of the extraction and quantitative study by liquid chromatography/ atmospheric pressure ionization tandem mass spectrometry. *J. Agr. Food Chem.* 2004, *53*, 424-429.

Camel, V. Recent extraction techniques for solid matrices-supercritical fluid extraction, pressurized fluid extraction and microwave-assisted extraction: their potential and pitfalls. *Analyst* 2001, *126*, 1182-1193.

Caridi, D.; Trenerry, V. C.; Rochfort, S.; Duong, S.; Laugher, D.; Jones, R. Profiling and quantifying quercetin glucosides in onion (*Allium cepa* L.) varieties using capillary zone electrophoresis and high performance liquid chromatography. *Food Chem.* 2007, *105*, 691-699.

Carrillo, L. C.; Londoño-Londoño, J.; Gil, A. Comparison of polyphenol, methylxanthines and antioxidant activity in *Theobroma cacao* beans from different cocoa-growing areas in Colombia. *Food Res. Int.* 2014, *60*, 273-280.

Chankvetadze, B. *Capillary Electrophoresis in Chiral Analysis*; John Wiley & Sons: Chichester, 1997; pp 251-255.

Chirinos, R.; Rogez, H.; Campos, D.; Pedreschi, R.; Larondelle, Y. Optimization of extraction conditions of antioxidant phenolic compounds from mashua (*Tropaeolum tuberosum* Ruíz & Pavón) tubers. *Sep. Purif. Technol.* 2007, *55*, 217-225.

Clapperton, J.; Lockwood, R.; Romanczyk, L.; Hammerstone, J. F. Contribution of genotype to cocoa (*Theobroma cacao* L.) flavour. *Trop. Agr.* (Trinidad) 1994, *71*, 303-308.

Cooper, K. A.; Campos-Giménez, E.; Jiménez Álvarez, D.; Nagy, K. L.; Donovan, J. L.; Williamson, G. Rapid reversed phase ultra-performance liquid chromatography analysis of the major cocoa polyphenols and inter-relationships of their concentrations in chocolate. *J. Agr. Food Chem.* 2007, *55*, 2841-2847.

Counet, C.; Collin, S. Effect of the number of flavanol units on the antioxidant activity of procyanidin fractions isolated from chocolate. *J. Agr. Food Chem.* 2003, *51*, 6816-6822.

Counet, C.; Callemien, D.; Collin, S. Chocolate and cocoa: New sources of *trans*-resveratrol and *trans*-piceid. *Food Chem.* 2006, *98*, 649-657.

Delcour, J. A.; Janssens de Varebeke, D. A new colorimetric assay for flavonoids in pilsner beers. *J. I. Brewing* 1985, *91*, 37-40.

Donovan, J. L.; Crespy, V.; Oliveira, M.; Cooper K. A.; Gibson, B. B.; Williamson, G. (+)-Catechin is more bioavailable than (–)-catechin: Relevance to the bioavailability of catechin from cocoa. *Free Radical Res.* 2006, *40*, 1029-1034.

Ellis, C. J.; Yeap Foo L.; Porter, L. J. Enantiomerism: a characteristic of the proanthocyanidin chemistry of the *Monocotyledonae*. *Phytochemistry* 1983, *22*, 483-487.

Ford, B. L.; Bai, J.; Manthey, J.; Baldwin, E. A. Improved removal of ascorbate interference in the Folin-Ciocalteu assay of "Total Phenolic Content." *Proc. Fla. State Hortic. Soc.* 2010, *123*, 220-222.

Georgé, S.; Brat, P.; Alter, P.; Amiot, M. J. Rapid Determination of Polyphenols and Vitamin C in Plant-Derived Products. *J. Agr. Food Chem.* 2005, *53*, 1370-1373.

Gotti, R.; Fiori, J.; Mancini, F.; Cavrini, V. Modified micellar electrokinetic chromatography in the analysis of catechins and xanthines in chocolate. *Electrophoresis* 2004, *25*, 3282-3291.

Gotti, R.; Furlanetto, S.; Pinzauti, S.; Cavrini, V. Analysis of catechins in *Theobroma cacao* beans by cyclodextrin-modified micellar electrokinetic chromatography. *J. Chromatogr.* A 2006, *1112*, 345-352.

Gu, L. W.; House, S. E.; Wu, X. L.; Ou, B. X.; Prior, R.L. Procyanidin and catechin contents and antioxidant capacity of cocoa and chocolate products. *J. Agr. Food Chem.* 2006, *54*, 4057-4061.

Guyot, S.; Marnet, N.; Laraba, D.; Sanoner, P.; Drilleau, J. F. Reversed-phase HPLC thiolysis for quantitative estimation and characterization of the four main classes of phenolic compounds in different tissues zones of French cider apple variety (*Malus domestica var. Kermerrien*). *J. Agr. Food Chem.* 1998, *46*, 1698-1705.

Hammerstone, J. F.; Lazarus, S. A.; Mitchell, A. E.; Rucker, R.; Schmitz, H. H. (1999). Identification of procyanidins in cocoa (*Theobroma cacao*) and chocolate using high-performance liquid chromatography/mass spectrometry. *J. Agr. Food Chem.* 1999, *47*, 490-496.

Hu, Y.; Pan, Z. J.; Liao, W.; Li, J.; Gruget, P.; Kitts, D. D.; Lu, X. Determination of antioxidant capacity and phenolic content of chocolate by attenuated total reflectance-Fourier transformed-infrared spectroscopy. *Food Chem.* 2016, *202*, 254-261.

Ignat, I.; Volf, I.; Popa, V. I. A critical review of methods for characterisation of polyphenolic compounds in fruits and vegetables. *Food Chem.* 2011, *126*, 1821-1835.

Isabelle, M.; Lee, B. L.; Lim, M. T.; Koh, W. P.; Huang, D.; Ong, C. N. Antioxidant activity and profiles of common vegetables in Singapore. *Food Chem.* 2010, *120*, 993-1003.

Jonfia-Essien, W. A.; West, G.; Alderson, P. G.; Tucker, G. Phenolic content and antioxidant capacity of hybrid variety cocoa beans. *Anal. Methods* 2008, *108*, 1155-1159.

Kealey, K. S.; Snyder, R. M.; Romanczyk, L. J.; Geyer, H. M.; Myers, M. E.; Withcare, E. J.; Hammerstone, J. F.; Schmitz, H. H. Cocoa components, edible products having enhanced polyphenol content, methods of making same and medical uses. *Patent Cooperation Treaty (PCT) WO 98/09533*, Mars Incorporated, USA, 1998.

Kofink, M.; Papagiannopoulos, M.; Galensa, R. (−)-Catechin in cocoa and chocolate: occurrence and analysis of an atypical flavan-3-ol enantiomer. *Molecules* 2007, *12*, 1274-1288.

Kothe, L.; Zimmermann, B. F.; Galensa, R. Temperature influences epimerization and composition of flavanol monomers, dimers and trimmers during cocoa bean roasting. *Food Chem.* 2013, *141*, 3656-3663.

Kramling, T. E.; Singleton, V. E. An estimate of the nonflavonoid phenols in wines. *Am. J. Enol. Viticult.* 1969, *20*, 86-92.

Kyi, T. M.; Daud, W. R. W.; Mohammad, A. B.; Wahid Samsudin, M.; Kadhum, A. A. H.; Talib, M. Z. M. The kinetics of polyphenol degradation during the drying of Malaysian cocoa beans. *Int. J. Food Sci. Technol.* 2005, *40*, 323-331.

Lee, K. W.; Kim, Y. J.; Lee, H. J.; Lee, C. Y. Cocoa has more phenolic phytochemicals and a higher antioxidant capacity than teas and red wine. *J. Agr. Food Chem.* 2003, *51*, 7292-7295.

Lucci, P.; Saurina, J.; Núñez, O. Trends in LC-MS and LC-HRMS analysis and characterization of polyphenols in food. *TrAC* 2017, *88*, 1-24.

Machonis, P. R.; Jones, M. A.; Schoneberg, B. T.; Kwik-Uribe, C. Method for the determination of catechin and epicatechin enantiomers in cocoa-based ingredients and products by high-performance liquid chromatography: Single laboratory validation. *J. AOAC Int.* 2012, *95*, 500-507.

Markham, K. R.; Bloor, S. J. Analysis and identification of flavonoids in practice. In *Flavonoids in health and disease*; Rice-Evans, C.A., Packer, L., Eds.; Marcel Dekker Inc.: New York, Basel, Hong Kong, 1998. pp 1-33.

Miller, K. B.; Hurst, W. J.; Payne, M. J.; Stuart, D. A.; Apgar, J.; Sweigart, D. S.; Ou, B. Impact of alkalization on the antioxidant and flavanol content of commercial cocoa powders. *J. Agr. Food Chem.* 2008, *56*, 8527-8533.

Motilva, M. J.; Serra, A.; Macia, A. Analysis of food polyphenols by ultra-high performance liquid chromatography coupled to mass spectrometry: an overview, *J. Chromatogr. A* 2013, *1292*, 66-82.

Natsume, M.; Osakabe, N.; Yamagishi, M.; Takizawa, T.; Nakamura, T.; Miyatake, H.; Hatano, T.; Yoshida, T. Analyses of phenols in cacao liquor, cocoa, and chocolate by normal-phase and reversed phase HPLC. *Biosci. Biotechnol. Biochem.* 2000, *64*, 2581-2587.

Niemenak, N.; Rohsius, C.; Elwers, S.; Ndoumou, D. O.; Lieberei, R. Comparative study of different cocoa (*Theobroma cacao* L.) clones in terms of their phenolics and anthocyanins contents. *J. Food Compos. Anal.* 2006, *19*, 612-619.

Oracz, J.; Nebesny, E.; Żyżelewicz, D. Changes in the flavan-3-ols, anthocyanins, and flavanols composition of cocoa beans of different *Theobroma cacao* L. groups affected by roasting conditions. *Eur. Food Res. Technol.* 2015, *241*, 663-681.

Ortega, N.; Romero, M. P.; Macià, A.; Reguant, J.; Anglès, N.; Morelló, J. R.; Motilva, M. J. Obtention and characterization of phenolic extracts from different cocoa sources. *J. Agr. Food Chem.* 2008, *56*, 9621-9627.

Ortega, N.; Romero, M. P.; Macià, A.; Reguant, J.; Anglès, N.; Morelló, J. R.; Motilva, M. J. Comparative study of UPLC-MS/MS and HPLC-MS/MS to determine procyanidins and alkaloids in cocoa samples. *J. Food Compos. Anal.* 2010, *23*, 298-305.

Ough, C. S.; Amerine, M. A. *Methods for analysis of must and wines*. 2th ed.; John Wiley & Sons: New York, 1988.

Patras, M. A.; Milev, B. P.; Vrancken, G.; Kuhnert, N. Identification of novel cocoa flavonoids from raw fermented cocoa beans by HPLC-MS^n. *Food Res. Int.* 2014, *63*, 353-359.

Pedan, V.; Fischer, N.; Rohn, S. Extraction of cocoa proanthocyanidins and their fractionation by sequential centrifugal partition chromatography and gel permeation chromatography. *Anal. Bioanal. Chem.* 2016, *408*, 5905-5914.

Pękal, A.; Pyrzynska, K. Evaluation of aluminium complexation reaction for flavonoid content assay. *Food Anal. Method*. 2014, *7*, 1776-1782.

Plaza, M.; Oliveira, D.; Nilsson, A.; Turner, C. Green and efficient extraction method to determine polyphenols in cocoa and cocoa products. *Food Anal. Method* **2017**, *10*, 2677-2691.

Porter, L. J.; Hrstich, L. N.; Chan, B. G. The conversion of procyanidins and prodelphinidins to cyanidin and delphinidin. *Phytochemistry* 1986, *25*, 223-230.

Porter, L. J.; Ma, Z.; Chan, B. G. Cocoa procyanidins: major flavonoids and identification of some minor metabolites. *Phytochemistry* 1991, *30*, 1657-1663.

Rigaud, J.; Escribano-Bailon, M.; Prieur, C.; Bouquet, J. M.; Cheynier, V. Normal phase high-performance liquid chromatography separation of procyanidins from cocoa beans and grape seeds. *J. Chromatogr. A* 1993, *654*, 255-260.

Romanczyk, L. J.; Hammerstone, J. F.; Buck, M. M.; Post, L. S.; Cipolla, G. G.; Micceland, C. A.; Mundt, J. A.; Schmitz, H. H. Cocoa extract compounds and methods for making and using the same. *Patent Cooperation Treaty (PCT) WO 97/36497*, Mars incorporated, USA, 1997.

Sanbongi, C.; Osakabe, N.; Natsume, M.; Takizawa, T.; Gomi, S.; Osawa, T. Antioxidative polyphenols isolated from *Theobroma cacao*. *J. Agr. Food Chem.* 1998, *46*, 454-457.

Sánchez-Rabaneda, F.; Jáuregui, O.; Casals, I.; Andrés-Lacueva, C.; Izquierdo-Pulido, M., Lamuela-Raventós, R. M. Liquid chromatographic/electrospray ionization mass spectrometry study of the phenolic composition of cocoa (*Theobroma cacao*). *J. Mass Spectrom.* 2003, *8*, 35-42.

Sánchez-Rangel, J. C.; Benavides, J.; Basilio Heredia, J.; Cisneros-Zevallos, L.; Jacobo-Velázquez, D. A. The Folin-Ciocalteu assay revisited: improvement of its specificity for total phenolic content determination. *Anal. Methods* 2013, *5*, 5990-5999.

Sarmento, L. A. V.; Machado, R. A. F.; Petrus, J. C. C.; Tamanini, T. R.; Bolzan, A. Extraction of polyphenols from cocoa seeds and concentration through polymeric membranes. *J. Supercrit. Fluid.* 2008, *45*, 64-69.

Schroeter, H.; Heiss, C.; Balzer, J.; Kleinbongard, P.; Keen, C. L.; Hollenber, N. K.; Sies, H.; Kwik-Uribe, C.; Schmitz, H. H.; Kelm, M. (−)-Epicatechin mediated beneficial effects of flavanol-rich cocoa on vascular function in human. *Proceedings of the National Academy of Sciences of the United States of America* 2006, *103*, 1024-1029.

Serra Bonvehí, J.; Ventura Coll, F. Evaluation of bitterness and astringency of polyphenolic compounds in cocoa powder. *Food Chem.* 1997, *60*, 365-370.

Shahidi, F.; Naczk, M. *Food Phenolics: Sources, Chemistry, Effects, Applications*; Technomic Publishing Company Inc.: Lancaster, PA, USA, 1995; pp 281-319.

Singleton, V. L.; Orthofer, R.; Lamuela-Raventós, R. M. Analysis of total phenols and their oxidation substrates and antioxidants by means of Folin-Ciocalteu reagent. *Methods Enzymol.* 1999, *299*, 152-178.

Todorovic, V.; Redovnikovic, I. R.; Todorovic, Z.; Jankovic, G.; Dodevska, M.; Sobajic, S. Polyphenols, methylxanthines, and antioxidant capacity of chocolates produced in Serbia. *J. Food Compos. Anal.* 2015, *41*, 137-143.

Tomás-Barberán, F. A.; Cienfuegos-Jovellanos, E.; Marín, A.; Muguerza, B.; Gil-Izquierdo, A.; Cerdá, B.; Zafrilla, P.; Morillas, J.; Mulero, J.; Ibarra, A.; Pasamar, M. A.; Ramón, D.; Espín, J. C. A new process to develop a cocoa powder with higher flavonoid monomer content and enhanced bioavailability in healthy humans. *J. Agr. Food Chem.* 2007, *55*, 3926-3935.

Tsao, R.; Yang, R. Optimization of a new mobile phase to know the complex and real polyphenolic composition: towards a total phenolic index using high-performance liquid chromatography. *J. Chromatogr. A* 2003, *1018*, 29-40.

Vinson, J. A.; Proch, J.; Bose, P. Determination of quantity and quality of polyphenol antioxidants in foods and beverages. *Method. Enzymol.* 2001, *335*, 103-114.

Waterhouse, A. L.; Sirley, J. R.; Donovan, J. L. Antioxidants in chocolate. *Lancet* 1996, *348*, 834.

Wollgast, J.; Anklam, A. Review on polyphenols in *Theobroma cacao*: changes in composition during the manufacture of chocolate and methodology for identification and quantification. *Food Res. Int.* 2000, *33*, 423-447.

Wollgast, J.; Pallaroni, L.; Agazzi, M. E.; Anklam, E. Analysis of procyanidins in chocolate by reversed-phase high-performance liquid chromatography with electrospray ionization mass spectrometric and tandem mass spectrometric detection. *J. Chromatogr. A* 2001, *926*, 211-220.

In: The Diversified Benefits of Cocoa … ISBN: 978-1-53613-258-8
Editor: Bonifacia Zayas Espinal

Chapter 2

DARK CHOCOLATE INTAKE AND NONCOMMUNICABLE DISEASES

Abbe Maleyki Mhd Jalil*
School of Nutrition and Dietetics, Faculty of Health Sciences,
Universiti Sultan Zainal Abidin (UniSZA), Kuala Nerus,
Terengganu, Malaysia

ABSTRACT

About forty million people die annually as a consequence of noncommunicable diseases (NCDs), accounting for approximately 70% of all death globally. Cardiovascular diseases account for 17.7 million deaths annually, followed by cancers (8.8 million), respiratory diseases (3.9 million) and diabetes mellitus, which is responsible for 1.6 million deaths. Smoking, physical inactivity, unhealthy diet (eating out and consuming take-away foods, which are high in salt and low in fibre) and excessive alcohol intake have all been identified as established risk factors for NCDs. The preventive strategies for NCDs have been focused on these modifiable risk factors. The recommendation of healthy dietary practices such as eating a healthy breakfast, preparing home-made meals

* Corresponding Author Email: abbemaleyki@unisza.edu.my.

low in salt and fat, and eating sufficient quantities of fruits and vegetables, is one strategy used by health professionals to curb NCDs. Fruits and vegetables are rich in dietary fibre and also a good source of vitamins, minerals and (poly)phenols. Epidemiological studies have shown that the consumption of fruits and vegetables is negatively correlated with the incidence of NCDs. However, other foods such as dark chocolate contain higher concentrations of (poly)phenols per 100 g compared with fruits and vegetables. Dark chocolate contains an average of 1860 mg total (poly)phenols per 100 g fresh weight. Numerous studies have shown the diverse health benefits of dark chocolate, particularly against NCDs, which are due to its high (poly)phenol content. Hence, this chapter will examine the current research in relation to the benefits of dark chocolate vis-à-vis NCDs and the associated risk factors.

Keywords: dark chocolate, (poly)phenols, flavanols, cardiovascular diseases, T2DM

DARK CHOCOLATE

Chocolate manufacturing is a trade secret and practices vary between companies. In general, chocolate is made from a combination of cocoa mass (or cocoa powder), cocoa butter (cocoa fat), cocoa liquor, milk powder and sugar (Bywaters, 1930; Barry Callebaut, 2017). The ingredient compositions vary for white, milk, semi-sweet and dark chocolates. White chocolate is made from cocoa butter, sugar and milk powder; milk chocolate is composed of cocoa liquor, cocoa butter, sugar and milk powder; and dark chocolate is made with cocoa liquor, cocoa butter and sugar (Barry Callebaut, 2017). No milk powder is added in the production of dark chocolate, and no cocoa mass or cocoa liquor is added in the production of white chocolate. Hence, in terms of health benefits, we cannot simply classify white chocolate as a 'real' chocolate. Different types of chocolate are marketed, with a clear label on the packaging, as either white, semi-sweet, milk, or dark chocolate (see Figure 1).

Figure 1. Commercially available chocolates in the market (from the top down: white, dark, semi-sweet, and milk chocolates).

(POLY)PHENOLS IN DARK CHOCOLATE

(Poly)phenols are a diverse group of compounds found in the plant kingdom. (Poly)phenols are products of the secondary metabolism of plants, and provide protection against pathogens and parasites, as well as providing plants with their unique colour (Liu, 2004). They are abundantly present in foodstuffs, including whole grains (e.g., sorghum and barley in the range of 2.6 to 39.7 mg per 100 g fresh weight), fruits (e.g., apples, blackberries, cranberries, grapes, peaches, pears, plums, raspberries, and strawberries in the range of 8.6 to 496.0 mg per 100 g fresh weight), vegetables (e.g., cabbage, celery, red onion, and parsley in the range of 0.1 to 139.5 mg per 100 g fresh weight) and beverages (e.g., black tea, coffee, cocoa and red wine in the range of 0.89 to 477.5 mg per 100 g fresh weight) (Bravo et al., 1998; Chung et al., 1998; Duthie & Crozier, 2000; Neveu et al., 2010). Aside from the foods specified above, another significant source of (poly)phenols is chocolate. However, (poly)phenol content varies from white to milk and dark chocolates. The ratio of non-fat cocoa solids (NFCS) added during chocolate production is higher for dark chocolate, followed by semi-sweet and milk chocolates (Miller et al., 2006). ‘Baking’ chocolate contains the highest NFCS with 45–49%, followed by dark chocolate, semi-sweet chocolate and milk chocolate with 20–30%, 15–19%, and 5–7%, respectively (Miller et al., 2006). Higher levels of NFCS indicate higher levels of (poly)phenol content. The (poly)phenol content of different types of chocolates and cocoa beverages is set out in Table 1. Factors that result in (poly)phenols variability in the final product include the origin of the cocoa beans (Ghana vs Trinidad), primary and secondary processing conditions (e.g., the cocoa drying and fermentation process) and the process of chocolate making (Cooper et al., 2007; Azizah et al., 2007; Caligiani et al., 2007). Vinson et al., (1999) demonstrated that the highest (poly)phenol content can be found in cocoa powder, followed by dark chocolate, milk chocolate and hot chocolate (the beverage).

BIOAVAILABILITY OF COCOA (POLY)PHENOLS

Before cocoa (poly)phenols can exert any health benefits or reach the target tissue, it must be made available in the gastrointestinal tract. The study of bioavailability is important to determining how (poly)phenols are metabolised in the gut, appear in the plasma, and are degraded in the large intestine for the production of phenolic acids. (Poly)phenol availability can be quantified by measuring the concentration of (poly)phenols in the plasma and urine after the ingestion of a known quantity of foodstuffs or pure (poly)phenolic compounds (Scalbert & Williamson, 2000). The bioavailability of (poly)phenols greatly depends on their chemical structure, glycosylation, acylation, conjugation and polymerisation (Scalbert & Williamson, 2000).

Table 1. (Poly)phenol content in cocoa and cocoa products

Cocoa products	(Poly)phenol content	Type of polyphenols	References
Dark chocolate	1860 mg/ 100 g fresh weight	(+)-catechin, (+)-gallocatechin, (-)-epicatechin 3-O-gallate, (-)-epicatechin-(2a-7)(4a-8)-epicatechin 3-O-galactoside, (-)-epigallocatechin, (-)-epigallocatechin 3-O-gallate, procyanidin B1, procyanidin trimer C1, trimer-decamer*, quercetin, ferulic acid, resveratrol	Mathur et al., (2002); Arts et al., (1999); Vinson et al., (1999); Rothwell et al., (2013)
Milk chocolate	854 mg/ 100 g fresh weight	(+)-catechin, (+)-gallocatechin, (-)-epicatechin, (-)-epicatechin 3-O-gallate, (-)-epigallocatechin, (-)-epigallocatechin 3-O-gallate, trimer-decamer*	Adamson et al., (1999); Vinson et al., (1999); Rothwell et al., (2013)
Cocoa powder	5824 mg/ 100 g fresh weight	(+)-catechin, (-)-epicatechin, (-)-epicatechin-(2a-7)(4a-8)-epicatechin 3-O-galactoside, cinnamtannin A2, procyanidin dimer B1, procyanidin dimer B2, procyanidin trimer C1, quercetin 3-O-arabinoside, 3-methylcatechol, 4-ethylcatechol, 4-methylcatechol, vanillin, mellein, catechol, phenol, pyrogallol, benzoic acid, caffeoyl aspartic acid, p-coumaroyl tyrosine, matairesinol, secoisolariciresinol, 4-hydroxybenzoic acid, gallic acid, protocatechuic acid, syringic acid, vanillic acid, caffeic acid, cinnamic acid, ferulic acid, p-coumaric acid, sinapic acid	Vinson et al., (1999); Maleyki et al., (2010); Rothwell et al., (2013)
Cocoa beverage	7.21 mg/ 200 ml	(+)-catechin, (-)-epicatechin, procyanidin B2, procyanidin B5, procyanidin C1	Mathur et al., (2002); Rios et al., (2002); Maleyki et al., (2007)

Note: * Trimer-decamer: polymerised (poly)phenol consists of 3-9 units of monomers.

Different sizes of (poly)phenols may be absorbed differently. For instance, smaller monomeric flavonoids (such as catechin and epicatechin) are absorbed in the small intestine, but polymeric procyanidins (such as procyanidin B1 and procyanidin trimer C1) may be metabolised by colonic microflora, followed by either absorption of the metabolites or excretion in the faeces. After absorption, the monomers and dimers may be methylated, sulphated, or glucuronidated in the liver (Scalbert & Williamson, 2000; Henning et al., 2008). The majority of (poly)phenols, with the exception of catechin and epicatechin, present as glycosides (attached to a sugar moiety). The absorption of these glycosides is dependent on the position of sugar linkage (Depeint et al., 2002). Glycosidic flavonoids require hydrolysis by colonic microflora to cleave the sugar and release the aglycone for absorption, whereas aglycones without a sugar moiety can passively diffuse through the small intestine.

Holt et al., demonstrated that plasma procyanidin dimer, (–)-epicatechin and (+)-catechin was detected as early as 0.5 h and reached maximal concentrations by 2 h following acute intake of flavanol-rich cocoa (Holt et al., 2002). Another study demonstrated an increased in 2 h plasma epicatechin with a peak concentration of 21.2 nmol/l following administration of chocolate containing 148 mg of procyanidins compared with baseline (0 h) (Schramm et al., 2001). Similarly, plasma (–)-epicatechin was detected between 0.5 and 1 h as sulphated, glucuronidated, or methylated forms following chocolate consumption (Baba et al., 2000). Richelle et al., demonstrated that consumption of 80 g of chocolate containing 165 mg of epicatechin increased plasma levels of total free and conjugated epicatechin by 0.7 μmol/L (Richelle et al., 1999). Plasma concentration of procyanidin B2 increased and reached peak plasma level at 30 min, gradually decreasing thereafter until 300 min when supplemented with cocoa procyanidin B2 (epicatechin-(4β-8)- epicatechin) (Baba et al., 2002). Da Silva et al., demonstrated the presence of (–)-epicatechin in plasma as glucuronide and sulfate-glucuronide conjugated (free and O-methylated) forms (Da Silva et al., 1998). The glucuronidation of (–)-epicatechin occurs at position 3' of the B ring in humans (Natsume et al., 2003). This glucuronidation compound exhibits low antioxidant activity when compared to intact compounds, as it has lost the catechol structure of the B ring responsible for the antioxidative effects.

In addition to the chemical structures of (poly)phenols, the presence of protein and the interactions between them may directly or indirectly affect the bioavailability of (poly)phenols. Protein is one of the macronutrients that has been widely studied due to its interaction with (poly)phenols. The presence of protein in food matrices forms complexes with (poly)phenols (Ferruzzi et al., 2006). However, the evidence for the interference of protein with (poly)phenols remains questionable. One study showed that the addition of milk reduces antioxidant capacity by 30% (Tabernero et al., 2006). In contrast, Schroeter et al., found that the bioavailability of (poly)phenol monomers (such as epicatechin) remained unaffected when cocoa was ingested with milk (Schroeter et al., 2003). Both salivary protein and protein in the digestive tract (enzymes) could also influence the bioavailability of (poly)phenols (Baxter et al., 1997; Brunet et al., 2002).

Other physiological factors, such as stomach pH, might also have an effect on (poly)phenols' stability and bioavailability. Procyanidins dimers are stable in the stomach and thus available for absorption and metabolism (Rios et al., 2002). Unabsorbed (poly)phenols will be further metabolised in the large intestine into phenolic acids. High molecular weight (poly)phenols such as proanthocyanidins and theaflavins are metabolised in the large intestine by the gut microbiota into phenolic acids (Chen et al, 2012; Manach et al., 2004; Deprez et al., 2000). For instance, proanthocyanidins are degraded into a range of low molecular weight phenolic acids, namely, 2-(*p*-hydroxyphenyl)acetic acid, 2-(*p*-hydroxy-phenyl)propionic acid, 2-(*m*-hydroxyphenyl)acetic acid, 2-(*m*-hydroxy-phenyl)propionic acid, 5-(*m*-hydroxyphenyl)valeric acid, and phenyl-propionic acid (Deprez et al., 2000). Similarly, cocoa procyanidins result in the production of *m*-hydroxyphenylpropionic acid, ferulic acid, 3,4-dihydroxyphenylacetic acid, *m*-hydroxyphenylacetic acid, vanillic acid and *m*-hydroxybenzoic acid in urine (Rios et al., 2002). Thus, it is clear that (poly)phenols are bioavailable but this bioavailability is dependent on their chemical structures as well as the physiological conditions in the gastrointestinal tract. Other factors such as the food matrix (chocolate drinks vs chocolate bars) and texture (hard vs soft) in addition to the presence of other food components such as dietary fibre might also affect (poly)phenols' bioavailability. For instance, eating a chocolate bar containing (poly)phenols and soluble dietary fibre (e.g., β-glucan) is different from eating a chocolate bar separately from porridge containing β-glucan. Recently, Jalil et al., demonstrated that the (poly)phenols in black tea could form complexes with dietary fibre (β-glucan), which in turn reduce *in vitro* digestibility (Jalil et al., 2015). This is an area of interest and warrants further investigation.

Dark Chocolate and Cardiovascular Diseases

Recent prospective studies demonstrated the relationship between chocolate consumption and coronary heart disease (CHD) (Yuan et al., 2017). Yuan et al., conducted a meta-analysis of the association between chocolate consumption and CHD among the populations of Europe, Australia and the United States of America (Yuan et al., 2017). The prospective studies were adjusted for factors such as age, sex, physical activity, alcohol intake, BMI, dietary energy intake, waist circumference, blood pressure, cholesterol level and socioeconomic status, among others (Buijsse et a., 2010; Janszky et al., 2009; Kwok et al., 2015; Larsson et al., 2016; Lewis et al., 2010; Mink et al., 2007). The follow-up periods ranged from 8.1 to 16 years, and the relative risk (RR) was calculated across chocolate intake of one, three, seven and ten servings per week. Yuan et al., concluded that chocolate consumption of three or more servings per week had little effect in reducing the risk of CHD (Yuan et al., 2017). A food frequency questionnaire was used to assess the frequency of chocolate intake; however, no data is available about the type of chocolate consumed. As discussed above, dark chocolate contains higher levels of (poly)phenols, and can therefore be expected to hold greater health benefits than milk chocolate. This could be investigated through a specific randomised controlled study.

The progression from normal to atherosclerotic tissues is due to an increase in oxidised LDL, overproduction of nitric-oxide, and inflammation at the cellular and molecular levels. LDL oxidation has been used as a biomarker to study lipid oxidation because it is hypothesised that oxidised LDL contributes to atherosclerosis. In addition, another assays such as cholesteryl ester hydroperoxides, phosphatidyl-choline hydro-peroxides, malondialdehyde (MDA), and F_2-isoprostanes have also been used to determine LDL oxidation. Among these assays, F_2-isoprostanes have received major attention because they are the product of arachidonic acid oxidation, which is considered the most reliable indicator of lipid peroxidation (Rietveld & Wiseman, 2002).

Studies have demonstrated that cocoa (poly)phenols ameliorate LDL oxidation (Kondo et al., 1996; Pearson et al., 2001). Kondo et al., demonstrated that cocoa powder extract prolongs the lag time of LDL oxidation in a concentration-dependent manner (Kondo et al., 1996). It has been proposed that cocoa (poly)phenols, specifically catechins, may prevent LDL oxidation either through free radical-quenching or metal-chelating abilities (Rice-Evans et al., 1996; Wiseman et al., 1997). Mathur et al., revealed that cocoa product supplementation significantly reduced LDL oxidation as shown by a longer lag time (9.8%) of conjugated diene formation compared to baseline (Mathur et al., 2002). Similarly, the lag time of LDL oxidation was 8% greater after cocoa powder and dark chocolate consumption compared with an American diet (Wan et al., 2001). Cocoa powder and dark chocolate supplementation also reduced the rate of oxidation and conjugated diene formation by 6.6% and 1.5%, respectively (Wan et al., 2001). Mechanistically, (poly)phenols directly inhibit peroxynitrite-induced LDL oxidation (Nijveldt et al., 2001). Nitric oxide (NO) is produced by nitric-oxide synthase in macrophages and is important for maintaining the dilation of blood vessels (Huk et al., 1998). However, peroxynitrite is formed when excess nitric oxide reacts with free radicals. The presence of other biomarkers such as endothelial function is recognised as an established risk factor for early events in cardiovascular disease (Fathi et al., 2004).

Endothelial function is measured as flow-mediated dilation (FMD) in coronary arteries and in the periphery as assessed by vasomotor function after an intra-arterial infusion of drugs that enhance the release of endothelial NO (Raitakari & Celermajer, 2000). Alternatively, a non-invasive ultrasound-based method can be used to measure FMD. Katz et al., extensively reviewed 17 interventional studies on the cardioprotective effects of dark chocolate on FMD (Katz et al., 2011). Improved FMD was observed in five studies among overweight adults, heart transplant recipients and smokers as well as healthy men following a single dosage (doses ranged from 176 to 805 mg) of either high-flavanol cocoa or dark chocolate (Faridi et al., 2008; Heiss et al., 2005; Flammer et al., 2007; Hermann et al., 2006; Schroeter et al., 2006). Sugared and non-sugared cocoa were equally effective at improving FMD. Interventional studies of longer duration (ranging from two days to six weeks) using a dose range of between 205 mg and 1008 mg total (poly)phenols showed an overall improvement in FMD compared with either low-flavanol or flavanol-free chocolate (Heiss et al., 2003; Fisher et al., 2003; Engler et al., 2004; Grassi et al., 2005; Fisher et al., 2006; Farouque et al., 2006; Balzer et al., 2008; Grassi et al., 2008; Njike et al., 2009; Shiina et al., 2009; Heiss et al., 2010). A review study concluded that a minimum of 2 g and a maximum of 5 g of cocoa intake is required to effect an increase in FMD in a healthy population (Monahan, 2012).

DARK CHOCOLATE AND DIABETES

Diabetes mellitus has become a major health problem, affecting more than 422 million people worldwide in 2014 (WHO, 2017). Globally, the total number of people with diabetes is likely to climb to 366 million by 2030, as estimated by the World Health Organization (WHO). In general, genetics, a sedentary lifestyle, abnormal glucose regulation, oxidative stress and obesity are major risk factors for diabetes. There is growing evidence suggesting that hyperglycaemia could excessively produce reactive oxygen species (ROS), which leads to oxidative stress at the cellular level (Rochette et al., 2014). Oxidative stress is increased in both types of diabetes mellitus (Types 1 and 2), but is more pronounced in Type 2 (T2DM) (Abou-Seif & Youssef, 2004). A clinical study demonstrated that treatment with antihyperglycaemic medication improves hyperglycaemia (Acharya et al., 2014). These therapeutic agents could modulate the signalling pathways associated with oxidative stress, thereby acting as indirect antioxidants in diabetic patients (Rochette et al., 2014). This improvement is associated with lower oxidative stress scores calculated based on antioxidant enzymes and damage markers. In addition to these drugs, bioactive food components, especially (poly)phenols, are important in ameliorating hyperglycaemia-related oxidative stress. The results of the Atherosclerosis Risk in Communities (ARIC) Cohort study showed that participants with a long-term chocolate intake of between two and six servings (up to 1 ounce) per week had a reduced diabetes risk compared with those who consumed less than 1 ounce of chocolate per month (Greenberg, 2015). Chocolate intake was measured using a food-frequency questionnaire and there is no indication of the type of chocolate being consumed by the population. Hence, additional prospective studies are warranted to investigate the association between dark chocolate intake and diabetes risk.

The consumption of cocoa containing 2.17 mg epicatechin, 1.52 mg catechin, 0.25 mg dimer and 0.13 mg trimer/g of cocoa extract for four weeks significantly reduced plasma oxidative stress (8-isprostane) in obese-diabetic rats (Abbe Maleyki et al., 2009). Moreover, endogenous antioxidant enzymes play a vital role in the regulation of cellular oxidative stress. To date, at least three types of antioxidant enzymes have been identified in the biological system. In general, they can be identified as superoxide dismutase (SOD), glutathione peroxidase (GPx), and catalase (Cat). Abbe Maleyki et al., demonstrated that (poly)phenol-rich cocoa supplementation increased SOD activity in obese-diabetic rats (Abbe Maleyki et al., 2009). Ioannone et al., established that dark chocolate reduced *in vitro* oxidative polymorphonuclear leukocytes (PMA)-induced compared with non-oxidised PMA (Ioannone et al., 2017). In humans, Ramirez-Sanchez et al., demonstrated that an intake of 18 g of dark chocolate containing 100 mg (−)-epicatechin for three months increased SOD and Cat protein expression and activity levels in patients with heart failure and T2DM (Ramirez-Sanchez et al., 2013). Although the patient sample is small (n = 5), this study provides robust data on the underlying mechanisms governing the relationship between cocoa (poly)phenols and skeletal muscle oxidative stress.

The use of oral hypoglycaemic drugs is the standard approach for controlling blood glucose response in diabetic patients (Saisho, 2015; Ho et al., 2011). Impaired glucose tolerance (IGT) and impaired fasting glucose (IFG) are recognised risk factors for the development of T2DM. 'IGT' is not a clinical term but rather a risk factor for future diabetes and/or adverse outcomes. This stage is also called the pre-diabetes stage and is considered a risk factor for the development of diabetes (NIDDK, 2017). A pre-diabetic subject is more likely to develop T2DM within 10 years and also has a higher risk of developing heart disease (NIDDK, 2017). Hence, from this point of view, glucose control is important in those with diabetes or with impaired glucose tolerance. Apart from therapeutic strategies, (poly)phenols have been shown to reduce glucose response (Deng et al., 2015; Bryans et al., 2007). Basu et al., demonstrated that a cocoa beverage containing 900 mg of (poly)phenols increased plasma insulin but had no effect on plasma glucose in diabetic subjects (n = 18) (Basu et al., 2015). The cocoa beverage was supplemented acutely after the consumption of a high-fat fast-food-style breakfast and hence may not have been able to reduce patients' glucose response. The evidence showed that fats or lipids in a food could lower the glycaemic response due to slower gastric emptying (Bell et al., 2015; Stacher et al., 1991). Another study by Mellor et al., demonstrated that the intake of 15 g of chocolate (three times per day, for a total of 45 g) containing 50 mg epicatechin for eight weeks did not reduce glycaemic response but increased HDL cholesterol in T2DM subjects (n = 12) (Mellor et al., 2013). Almoosawi et al., showed that dark chocolate supplementation (20 g) containing 500 mg of (poly)phenols reduced plasma glucose in overweight and obese subjects (n = 42) for four weeks compared with placebo (Almoosawi et al., 2012). The cocoa dose used by Almoosawi et al., was somewhat lower than that of Mellor et al., (20 g vs 45 g, respectively) but contained higher levels of (poly)phenols (500 mg vs 50 mg, respectively). In a larger sample of 47 T2DM subjects, chronic (one-year) chocolate supplementation containing 850 mg of (poly)phenols had no effect on glucose levels, but significantly reduced insulin levels (Curtis et al., 2012). The supplementation also improved HDL-c and decreased LDL-c concentrations.

Dark chocolate consumption for 56 days reduced fasting glucose levels without changes in insulin, compared with white chocolate, in T2DM subjects (Rostami et al., 2015). In this study, the (poly)phenols were calculated as total flavonoids (450 mg) and not as total (poly)phenols. The definition of the term total (poly)phenols might differ from that of total flavonoids, as the latter may only consider flavanols (e.g., catechin and epicatechin), and disregard other flavonoid sub-classes such as flavanol, flavone and flavanone. The sodium-dependent glucose transporter (SGLT1) is involved in glucose uptake from the apical surface of the intestinal lumen (follows concentration gradient of facilitated diffusion), while GLUT2 (a protein glucose carrier system) is involved in the exit of glucose (and also other sugars, e.g., fructose) to the basolateral surface (Roder et al., 2014). An *in vitro* study utilising rat intestine showed that 1 mM of epicatechin gallate (ECG) and epigallocatechin gallate (EGCG) reduced glucose uptake by 53 and 35%, respectively, when compared with the control (Kobayashi et al., 2000). The presence of the galloyl moiety on ECG and EGCG is important for blocking glucose uptake. Moreover, flavanols lack a carbon-carbon double bond between C2 and C3 of the C-ring, and hence have less electron density. An *in vitro* system demonstrated that the weaker electron density at the C-ring was associated with inhibitory activity of flavan-3-ol with salivary α-amylase (Bandyopadhyay et al., 2012).

Esser et al., supplemented overweight men with 70 g of dark chocolate containing high levels of flavanols (total flavanols = 1078 mg, including 349 mg of epicatechin) compared with normal flavanol-containing chocolate (total flavanols = 259 mg, including 97 mg of epicatechin) (Esser et al., 2014). The high–flavanol dark chocolate increased fasting plasma glucose compared with the normal-flavanol chocolate. Since both chocolates contained similar levels of cocoa (58% cocoa mass), theobromine and caffeine, the surprisingly high fasting plasma levels could be due to other reasons. Park et al., demonstrated that tea supplementation containing 500 mg epigallocatechin gallate (EGCG) reduced glucose levels at 30 and 60 min but increased glucose at 120 min by 29% when compared with the control (Park et al., 2009). The author suggested that green tea extract reduces early glucose (30 and 60 min) by decreasing the quantity of glucose entering the intestine, while the presence of gallate catechin in blood circulation produces a hyperglycaemic effect at 120 min by blocking glucose uptake into the tissues. The author's findings supported the results of an *in vitro* study that showed that cell lines treated with 1 and 10 μM EGCG (gallate catechin) decreased glucose (2-deoxy-[^{3}H]) uptake into hepatocytes (HepG2), adipocytes (3T3-L1), and beta cells (INS-1) (Park et al., 2009). However, there is no such evidence available for cocoa (poly)phenols. It seems that cocoa (poly)phenols might have a role in controlling blood glucose and also in reducing the risks associated with diabetes. However, it is important to note that the type of chocolate, dose, and the concentration of (poly)phenols vary from one study to the other.

CONCLUSION AND FUTURE DIRECTION

There is little evidence from existing prospective studies to conclude that moderate chocolate intake is associated with a reduced risk of CVD. One of the caveats is that the types of chocolate, whether milk, semi-sweet or dark, is not clearly defined in the prospective studies. However, it seems plausible to conclude from randomised controlled study that moderate chocolate intake reduces the risks associated with CVD. Dark chocolate intake of between 20 g and 70 g, containing between 50 mg and 1078 mg of either epicatechin or total flavanols, increased HDL-c and reduced glucose levels in T2DM subjects. Dark chocolate is a solid food and the (poly)phenols might be released too slowly in the gastrointestinal tract to confer any health benefits. Hence, dark chocolate with the same amount of (poly)phenols could be prepared as a beverage in combination with other food components such as dietary fibre. This will decrease luminal mixing and hence slow the blood glucose and insulin responses. Such an approach could be useful for the management of T2DM, but requires further investigation.

References

Abbe Maleyki, M. J.; Amin, I. (2010). *Journal of Food Biochemistry*. Antioxidant properties of cocoa powder, 34, 111–128.

Abbe Maleyki, M. J.; Amin, I.; Pei Pei, C.; Muhajir, H., & Syed Hasbullah, S. K. (2008). *Journal of Agricultural and Food Chemistry*. Effects of cocoa extract on glucometabolism, oxidative stress, and antioxidant enzymes in obese-diabetic rats, 56, 7877–7884.

Abbe Maleyki, M. J.; Amin, I.; Pei Pei, C.; Muhajir, H., & Syed Hasbullah, S. K. (2009). *Journal of the Science of Food and Agriculture*. Effects of cocoa extract containing polyphenols and methylxanthines on biochemical parameters of obese diabetic rats, 89, 130–137.

Abou-Seif, M. A.; Youssef, A. A. (2004). *Clinica Chimica Acta*. Evaluation of some biochemical changes in diabetic patients, 346, 161–170.

Acharya, J. D.; Pande, A. J.; Joshi, S. M.; Yajnik, C. S.; Ghaskadbi, S. S. (2014). *Diabetes/Metabolism Research and Reviews*. Treatment of hyperglycaemia in newly diagnosed diabetic patients is associated with a reduction in oxidative stress and improvement in β-cell function, 30, 590-598.

Adamson, G. E.; Lazarus, S. A.; Mitchell, A. E.; Prior, R. L.; Cao, G.; Jacobs, P. H.; Kremers, B. G.; Hammerstone, J. F.; Rucker, R. B.; Ritter, K. A.; Schmitz, H. H. (1999). *Journal of Agricultural and Food Chemistry*. HPLC method for the quantification of procyanidins in cocoa and chocolate samples and correlation to total antioxidant capacity, 47, 4184-4188.

Almoosawi, S.; Tsang, C.; Ostertag, L. M.; Fyfe, L.; Al-Dujaili, E. A. (2012). *Food & Function.* Differential effect of polyphenol-rich dark chocolate on biomarkers of glucose metabolism and cardiovascular risk factors in healthy, overweight and obese subjects: a randomized clinical trial, 3, 1035-1043.

Arts, I. C. W.; Hollman, P. C. H.; Kromhout, D. (1999). *The Lancet.* Chocolate as a source of tea flavonoids, 354, 488.

Azizah, O.; Amin, I.; Nawalyah, A. G.; Ilham, A. (2007). *Food Chemistry.* Antioxidant capacity and phenolic content of cocoa beans, 100, 1523-1530.

Baba, S.; Osakabe, N.; Yasuda, A.; Natsume, M.; Takizawa, T.; Nakamura, T.; Terao, J. (2000). *Free Radical Research*. Bioavailability of (-)-epicatechin upon intake of chocolate and cocoa in human volunteers, 33, 635-641.

Baba, S.; Osakabe, N.; Natsume, M.; Terao, J. (2002). *Free Radical Biology & Medicine*. Absorption and urinary excretion of procyanidin B2 [epicatechin-(4β-8)-epicatechin] in rats, 33, 142-148.

Balzer, J.; Rassaf, T.; Heiss, C.; Kleinbongard, P.; Lauer, T.; Merx, M.; Heussen, N.; Gross, H. B.; Keen, C. L.; Schroeter, H.; Kelm, M. (2008). *Journal of the American College of Cardiology.* Sustained benefits in vascular function through flavanol-containing cocoa in medicated diabetic patients a double-masked, randomized, controlled trial, 51, 2141–2149.

Barry Callebaut. *From cocoa to chocolate.* 2017 [2017 October 5th]. Available from: URL: https://www.barry-callebaut.com/stories/cocoa-chocolate.

Basu, A.; Betts, N. M.; Leyva, M. J.; Fu, D.; Aston, C. E.; Lyons, T. J. (2015). *Journal of Nutrition.* Acute Cocoa Supplementation Increases Postprandial HDL Cholesterol and Insulin in Obese Adults with Type 2 Diabetes after Consumption of a High-Fat Breakfast, 145, 2325-2332.

Baxter, N. J.; Lilley, T. H.; Haslam, E.; Williamson, M. P. (1997). *Biochemistry.* Multiple interactions between polyphenols and a salivary protein-rich protein result in complexation and precipitation, 36, 5566-5577.

Bell, K. J.; Smart, C. E.; Steil, G. M.; Brand-Miller, J. C.; King, B.; Wolpert, H. A. (2015). *Diabetes Care*. Impact of fat, protein, and glycemic index on postprandial glucose control in type 1 diabetes: implications for intensive diabetes management in the continuous glucose monitoring era, 38, 1008-1015.

Brand-Miller, J.; Holt, S. H. A.; De Jong, V.; Petocz, P. (2003). *Journal of Nutrition*. Cocoa powder increases postprandial insulinemia in lean young adults, 133, 3149-3152.

Bravo, L. (1998). *Nutrition Reviews*. Polyphenols: chemistry, dietary sources, metabolism, and nutritional significance, 56, 317-333.

Brunet, M. J.; Blade, C.; Salvado, M. J.; Arola, L. (2002). *Journal of Agricultural and Food Chemistry*. Human Apo A-I and rat transferrin are the principal plasma proteins that bind wine catechins, 50, 2708-2712.

Bryans, J. A.; Judd, P. A.; Ellis, P. R. (2007). *The Journal of the American College of Nutrition*. The effect of consuming instant black tea on postprandial plasma glucose and insulin concentrations in healthy humans, 26, 471-477.

Buijsse, B.; Weikert, C.; Drogan, D.; Bergmann, M.; Boeing, H. (2010). *European Heart Journal*. Chocolate consumption in relation to blood pressure and risk of cardiovascular disease in German adults, 31, 1616–1623.

Bywaters, H. W. (1930). *Journal of the Franklin Institute*. Modern methods of cocoa and chocolate manufacture, 209, 431-432.

Caligiani, A.; Cirlini, M.; Palla, G.; Ravaglia, R.; Arlorio, M. G. C. (2007). *Chirality*. MS detection of chiral markers in cocoa beans of different quality and geographic origin, 19, 329-334.

Chen, H.; Shurlknight, K.; Leung, T.; Sang, S. (2012). *Journal of Agricultural and Food Chemistry*. Structural identification of theaflavin trigallate and tetragallate from black tea using liquid chromatography/electrospray ionization tandem mass spectrometry, 60, 10850-10857.

Chung, K. T.; Wong, T. Y.; Wei, C. I.; Huang, Y. W.; Lin, Y. (1998). *Critical Reviews in Food Science and Nutrition*. Tannins and human health: a review, 38, 421-464.

Cooper, K. A.; Donovan, J. L.; Waterhouse, A. L.; Williamson, G. (2008). *British Journal of Nutrition*. Cocoa and health: a decade of research, 99, 1-11.

Cooper, K. A.; Campos-Gimenez, E.; Alvarez, D. J.; Nagy, K.; Donovan, J. L.; Williamson, G. (2007). *Journal of Agricultural and Food Chemistry*. Rapid reverse-phase ultra-performance liquid chromatography analysis of the major cocoa polyphenols and inter-relationship of their concentration in chocolate, 55, 2841-2847.

Curtis, P. J.; Sampson, M.; Potter, J.; Dhatariya, K.; Kroon, P. A.; Cassidy, A. (2012). *Diabetes Care*. Chronic ingestion of flavan-3-ols and isoflavones improves insulin sensitivity and lipoprotein status and attenuates estimated 10-year CVD risk in medicated postmenopausal women with type 2 diabetes: A 1-year, double-blind, randomized, controlled trial, 35, 226–232.

Da Silva, E. L.; Piskula, M.; Terao, J. (1998). *Free Radical Biology and Medicine*. Enhancement of antioxidative ability of rat plasma by oral administration of epicatechin, 24, 1209-1216.

Deng, Y. T.; Lin-Shiau, S. Y.; Shyur, L. F.; Lin, J. K. (2015). *Food & Function*. Pu-erh tea polysaccharides decrease blood sugar by inhibition of alpha-glucosidase activity *in vitro* and in mice, 6, 1539-1546.

Depeint, F.; Gee, J. M.; Williamson, G.; Johnson, I. T. (2002). *The Proceedings of the Nutrition Society*. Evidence for consistent patterns between flavonoid structures and cellular activities, 61, 97-103.

Deprez, S.; Brezillon, C.; Rabot, S.; Philippe, C.; Mila, I.; Lapierre, C.; Scalbert, A. (2000). *Journal of Nutrition*. Polymeric proanthocyanidins are catabolized by human colonic microflora into low-molecular-weight phenolic acids, 130, 2733-2738.

Duthie, G. & Crozier, A. (2000). *Current Opinion in Clinical Nutrition & Metabolic Care*. Plant-derived phenolic antioxidants, 3, 447-451.

Engler, M. B.; Engler, M. M.; Chen, C. Y.; Malloy, M. J.; Browne, A.; Chiu, E. Y.; Kwak, H. K.; Milbury, P.; Paul, S. M.; Blumberg, J.; Mietus-Snyder, M. L. (2004). *The Journal of the American College of Nutrition.* Flavonoid-rich dark chocolate improves endothelial function and increases plasma epicatechin concentrations in healthy adults, 23, 197–204.

Esser, D.; Mars, M.; Oosterink, E.; Stalmach, A.; Müller, M.; Afman. A. (2014). *The Official Journal of the Federation of American Societies for Experimental Biology.* Dark chocolate consumption improves leukocyte adhesion factors and vascular function in overweight men, 28, 1464–1473.

Faridi, Z.; Njike, V. Y.; Dutta, S.; Ali, A.; Katz, D. L. (2008). *The American Journal of Clinical Nutrition.* Acute dark chocolate and cocoa ingestion and endothelial function: a randomized controlled crossover trial, 88, 58–63.

Farouque, H. M.; Leung, M.; Hope, S. A.; Baldi, M.; Schechter, C.; Cameron, J. D.; Meredith, I. T. (2006). *Clinical Science.* Acute and chronic effects of flavanol-rich cocoa on vascular function in subjects with coronary artery disease: a randomized double-blind placebo-controlled study, 111, 71–80.

Fathi, R.; Haluska, B.; Isbel, N.; Short, L.; Marwick, T. H. (2004). *Journal of the American College of Cardiology.* The relative importance of vascular structure and function in predicting cardiovascular events, 43, 616–623.

Fisher, N. D.; Hughes, M.; Gerhard-Herman, M.; Hollenberg, N. K. (2003). *Journal of Hypertension.* Flavanol-rich cocoa induces nitric-oxide-dependent vasodilation in healthy humans, 21, 2281–2286.

Flammer, A. J.; Hermann, F.; Sudano, I.; Spieker, L.; Hermann, M.; Cooper, K. A.; Serafini, M.; Luscher, T. F.; Ruschitzka, F.; Noll, G.; Corti, R. (2007). *Circulation.* Dark chocolate improves coronary vasomotion and reduces platelet reactivity, 116, 2376–2382.

Grassi, D.; Desideri, G.; Necozione, S.; Lippi, C.; Casale, R.; Properzi, G.; Blumberg, J. B.; Ferri, C. (2008). *Journal of Nutrition.* Blood pressure is reduced and insulin sensitivity increased in glucose-intolerant, hypertensive subjects after 15 days of consuming high- polyphenol dark chocolate, 138, 1671–1676.

Grassi, D.; Lippi, C.; Necozione, S.; Desideri, G.; Ferri, C. (2005). *The American Journal of Clinical Nutrition.* Short-term administration of dark chocolate is followed by a significant increase in insulin sensitivity and a decrease in blood pressure in healthy persons, 81, 611-614.

Grassi, D.; Necozione, S.; Lippi, C.; Croce, G.; Valeri, L.; Pasqualetti, P.; Desideri, G.; Blumberg, J. B.; Ferri, C. (2005). *Hypertension.* Cocoa reduces blood pressure and insulin resistance and improves endothelium-dependent vasodilation in hypertensives, 46, 398-405.

Greenberg. J. A. (2015). *Clinical Nutrition.* Chocolate intake and diabetes risk, 34, 129-33.

Heiss, C.; Dejam, A.; Kleinbongard, P.; Schewe, T.; Sies, H.; Kelm, M. (2003). *The Journal of the American Medical Association.* Vascular effects of cocoa rich in flavan-3-ols, 290, 1030–1031.

Heiss, C.; Kleinbongard, P.; Dejam, A.; Perre, S.; Schroeter, H.; Sies, H.; Kelm, M. (2005). *Journal of the American College of Cardiology.* Acute consumption of flavanol-rich cocoa and the reversal of endothelial dysfunction in smokers, 46, 1276–1283.

Heiss, C.; Jahn, S.; Taylor, M.; Real, W. M.; Angeli, F. S.; Wong, M. L.; Amabile, N.; Prasad, M.; Rassaf, T.; Ottaviani, J. I.; Mihardja, S.; Keen, C. L.; Springer, M. L.; Boyle, A.; Grossman, W.; Glantz, S. A.; Schroeter, H.; Yeghiazarians, Y. (2010). *Journal of the American College of Cardiology.* Improvement of endothelial function with dietary flavanols is associated with mobilization of circulating angiogenic cells in patients with coronary artery disease, 56, 218–224.

Henning, S. M.; Choo, J. J.; Heber, D. (2008). *Journal of Nutrition.* Nongallated compared with gallated flavan-3-ols in green and black tea are more bioavailable, 138, 1529s-1534s.

Hermann, F.; Spieker, L. E.; Ruschitzka, F.; Sudano, I.; Hermann, M.; Binggeli, C.; Luscher, T. F.; Riesen, W.; Noll, G.; Corti, R. (2006). *Heart.* Dark chocolate improves endothelial and platelet function, 92, 119–120.

Ho, J.; Leung, A. K.; Rabi, D. (2011). *Recent Patents on Endocrine, Metabolic & Immune Drug Discovery.* Hypoglycemic agents in the management of type 2 diabetes mellitus, 5, 66-73.

Holt, R. R.; Lazarus, S. A.; Sullards, M. C.; Zhu, Q. Y.; Schramm, D. D.; Hammerstone, J. F.; Fraga, C. G.; Schmitz, H. H., Keen, C. L. (2002). *The American Journal of Clinical Nutrition.* Procyanidin dimmer B2 [epicatechin-(4β-8) - epicatechin] in human plasma after the consumption of a flavanol-rich cocoa, 76, 798-804.

Huk, I.; Brovkovych, V.; Nanobash, V. J. (1998). *British Journal of Surgery.* Bioflavonoid quercetin scavenges superoxide and increase nitric oxide concentrations in ischemic-reperfusion injury: an experimental study, 85, 1080-1085.

Ioannone, F.; Sacchetti, G.; Serafini, M. (2017). *Frontiers in Nutrition.* Effect of Dark chocolate extracts on Phorbol 12-Myristate 13-acetate-induced oxidative burst in leukocytes isolated by normo- weight and overweight/obese subjects, 4.

Jalil, A. M.; Edwards, C. A.; Combet, E.; Ibrahim, M.; Garcia, A. L. (2015). *International Journal of Food Sciences and Nutrition.* Combined effects of added beta glucan and black tea in breads on starch functionality, 66, 159-165.

Janszky, I.; Mukamal, K. J.; Ljung, R.; Ahnve, S.; Ahlbom, A.; Hallqvist, J. (2009). *Journal of Internal Medicine.* Chocolate consumption and mortality following a first acute myocardial infarction: The Stockholm Heart Epidemiology Program, 266, 248–257.

Katz, D. L.; Doughty, K.; Ali, A. (2011). *Antioxidants & Redox Signaling.* Cocoa and chocolate in human health and disease, 15, 2779-2811.

Kobayashi, Y.; Suzuki, M.; Satsu, H.; Arai, S.; Hara, Y.; Suzuki, K.; Miyamoto, Y. & Shimizu, M. (2000). *Journal of Agricultural and Food Chemistry.* Green tea polyphenols inhibit the sodium-dependent glucose transporter of intestinal epithelial cells by a competitive mechanism. 48, 5618-5623.

Kondo, K.; Hirano, R.; Matsumoto, A.; Igarashi, O.; Itakura, H. (1996). *The Lancet.* Inhibition of LDL oxidation by cocoa, 348, 1514.

Kwok, C. S.; Boekholdt, S. M.; Lentjes, M. A.; Loke, Y. K.; Luben, R. N.; Yeong, J. K.; Wareham, N. J.; Myint, P. K.; Khaw, K. T. (2015). *Heart.* Habitual chocolate consumption and risk of cardiovascular disease among healthy men and women, 101, 1279–1287.

Larsson, S. C.; Akesson, A.; Gigante, B.; Wolk. A. (2016). *Heart.* Chocolate consumption and risk of myocardial infarction: A prospective study and meta-analysis, 102, 1017–1022.

Lewis, J. R.; Prince, R. L.; Zhu, K.; Devine, A.; Thompson, P. L.; Hodgson, J. M. (2010). *Archives of Internal Medicine.* Habitual chocolate intake and vascular disease: A prospective study of clinical outcomes in older women, 170, 1857–1858.

Liu, R. H. (2004). *Journal of Nutrition.* Potential synergy of phytochemicals in cancer prevention: mechanism of action, 134, 3479S-3485S.

Manach, C.; Scalbert, A.; Morand, C.; Remesy, C.; Jimenez, L. (2004). *The American Journal of Clinical Nutrition*. Polyphenols: food sources and bioavailability, 79, 727-747.

Mathur, S.; Devaraj, S.; Grundy, S. M.; Jialal, I. (2002). *Journal of Nutrition.* Cocoa products decrease low density lipoprotein oxidative susceptibility but do not affect biomarkers of inflammation in humans, 132, 3663-3667.

Mellor, D. D.; Madden, L. A.; Smith, K. A.; Kilpatrick, E. S.; Atkin, S. L. (2013). *Diabetic Medicine.* High-polyphenol chocolate reduces endothelial dysfunction and oxidative stress during acute transient hyperglycaemia in type 2 diabetes: a pilot randomized controlled trial. Diabetic, 30, 478−483.

Miller, K. B.; Stuart, D. A.; Smith, N. L.; Lee, C. Y.; McHale, N. L.; Flanagan, J. A.; Ou, B.; Hurst, W. J. (2006). *Journal of Agricultural of Food Chemistry.* Antioxidant activity and polyphenol and procyanidin contents of selected commercially available cocoa-containing and chocolate products in the United States, 54, 4062-4068.

Mink, P. J.; Scrafford, C. G.; Barraj, L. M.; Harnack, L.; Hong, C. P.; Nettleton, J. A.; Jacobs, D. R., (2007). *The American Journal of Clinical Nutrition.* Flavonoid intake and cardiovascular disease mortality: A prospective study in postmenopausal women, 85, 895–909.

Monahan, K. D. (2012). *Archives of Biochemistry and Biophysics*. Effect of cocoa/chocolate ingestion on brachial artery flow-mediated dilation and its relevance to cardiovascular health and disease in humans, 527, 90–94.

Natsume, M.; Osakabe, N.; Oyama, M.; Sasaki, M.; Baba, S.; Nakamura, Y.; Osawa, T.; Terao, J. (2003). *Free Radical Biology & Medicine*. Structures of (-)-epicatechin gkucuronide identified from plasma and urine after oral ingestion of (-)-epicatechin: differences between human and rat, 34, 840-849.

Neveu, V.; Perez-Jimenez, J.; Vos, F.; Crespy, V.; Du Chaffaut, L.; Mennen, L.; Knox, C.; Eisner, R.; Cruz, J.; Wishart, D.; Scalbert, A. (2010). *Database: The Journal of Biological Databases and Curation*. Phenol-Explorer: an online comprehensive database on polyphenol contents in foods. Database (Oxford), 2010, bap024.

Nijveldt, R. J.; van Nood, E.; van Hoorn, D. E.; Boelens, P. G.; van Norren, K.; van Leeuwen, P. A. (2001). *American Journal of Clinical Nutrition*. Flavonoids: a review of probable mechanisms of action and potential applications, 74, 418-425.

National Institute of Diabetes and Digestive and Kidney Diseases (NIDDK). What is prediabetes? [2017 October 10th]. Available from: URL: https://www.niddk.nih.gov/health-information/diabetes/overview/what-is-diabetes/prediabetes-insulin-resistance.

Njike, V. Y.; Faridi, Z.; Shuval, K.; Dutta, S.; Kay, C D.; West, S. G.; Kris-Etherton, P. M.; Katz, D. L. (2011). *International Journal of Cardiology*. Effects of sugar-sweetened and sugar-free cocoa on endothelial function in overweight adults, 149, 83-88.

Park, J. H.; Jin, J. Y.; Baek, W. K.; Park, S. H.; Sung, H. Y.; Kim, Y. K.; Lee, J.; Song, D. K. (2009). *Journal of Physiology and Pharmacology*. Ambivalent role of gallated catechins in glucose tolerance in humans: a novel insight into non-absorbable gallated catechin-derived inhibitors of glucose absorption, 60, 101-109.

Pearson, D. A.; Schmitz, H. H.; Lazarus, S. A.; Keen, C. L. (2001). *Methods in Enzymology*. Inhibition of *in-vitro* low-density lipoprotein oxidation by oligomeric procyanidins present in chocolate and cocoas, 335, 350-360.

Raitakari, O. T.; & Celermajer D. S. (2000). *British Journal of Clinical Pharmacology*. Flow-mediated dilatation, 50, 397–404.

Ramirez-Sanchez, I.; Taub, P. R.; Ciaraldi, T. P.; Nogueira, L.; Coe, T.; Perkins, G.; Hogan, M.; Maisel, A. S.; Henry, R. R.; Ceballos, G.; Villarreal, F. (2013). *International Journal of Cardiology*. (−)-Epicatechin rich cocoa mediated modulation of oxidative stress regulators in skeletal muscle of heart failure and type 2 diabetes patients, 168, 3982–3990.

Rice-Evans, C. A.; Miller, N. J.; Paganga, G. (1996). *Free Radical Biological Medicine*. Structure-antioxidant activity relationships of flavonoids and phenolic acids, 20, 933-956.

Richelle, M.; Tavazzi, I.; Enslen, M.; Offord, E. (1999). *European Journal of Clinical Nutrition*. Plasma kinetics in man of epicatechin from black chocolate, 53, 22-26.

Rietveld, A. & Wiseman, S. (2003). *The American Society for Nutritional Sciences*. Antioxidant effects of tea: Evidence from human clinical trials, 133, 3285S-3292S.

Rios, L. Y.; Bennett, R. N.; Lazarus, S. A.; Remesy, C.; Scalbert, A.; Williamson, G. (2002). *The American Journal of Clinical Nutrition*. Cocoa procyanidins are stable during gastric transit in humans, 76, 1106-1110.

Rochette, L.; Zeller, M.; Cottin, Y.; Vergely, C. (2014). *Biochimica et Biophysica Acta*. Diabetes, oxidative stress and therapeutic strategies, 1840, 2709-2729.

Roder, P. V.; Geillinger, K. E.; Zietek, T. S.; Thorens, B.; Koepsell, H. & Daniel, H. (2014). *PLoS One*. The role of SGLT1 and GLUT2 in intestinal glucose transport and sensing, 9, e89977.

Rostami, A.; Khalili, M.; Haghighat, N.; Eghtesad, S.; Shidfar, F.; Heidari, I.; Ebrahimpour-Koujan, S.; Eghtesadi, M. (2015). *Atherosclerosis*.

High-cocoa polyphenol-rich chocolate improves blood pressure in patients with diabetes and hypertension, 11, 21–29.

Rothwell, J. A.; Pérez-Jiménez, J.; Neveu, V.; Medina-Ramon, A.; M'Hiri, N.; Garcia-Lobato, P.; Manach, C.; Knox, C.; Eisner, R.; Wishart, D. S.; Scalbert, A. (2013). *The Journal of Biological Databases and Curation.* Phenol-Explorer 3.0: a major update of the Phenol-Explorer database to incorporate data on the effects of food processing on polyphenol content, 2013.

Ruzaidi, A. M. M.; Amin, I.; Nawalyah, A. G.; Hamid, M.; Faizul, H. A. (2005). *Journal of Ethnopharmacology*. The effect of Malaysian cocoa extract on glucose levels and lipid profiles in diabetic rats, 98, 55-60.

Ruzaidi, A. M. M.; Abbe Maleyki, M. J.; Amin, I.; Nawalyah, A. G.; Muhajir, H. (2008). *Journal of the Science of Food and Agriculture.* Protective effect of polyphenol-rich extract prepared from Malaysian cocoa (Theobroma cacao) on glucose levels and lipid profiles in streptozotocin-induced diabetic rats, 88, 1442-1447.

Saisho, Y. (2015). *Endocrine, Metabolic & Immune Disorders – Drug Targets.* Metformin and Inflammation: Its Potential Beyond Glucose-lowering Effect, 15, 196-205.

Scalbert, A.; Williamson, G. (2000). *Journal of Nutrition.* Dietary intake and bioavailability of polyphenols, 130, 2073S-2085S.

Schramm, D. D.; Wang, J. F.; Holt, R. R.; Ensunsa, J. L.; Gonsalves, J. L.; Lazarus, S. A.; Schmitz, H. H.; German, J. B.; Keen, C. L. (2001). *The American Journal of Clinical Nutrition.* Chocolate procyanidins decrease the leukotriene-prostacyclin ratio in humans and human aortic endothelial cells, 73, 36-40.

Schroeter, H.; Holt, R. R.; Orozco, T. J.; Schmitz, H. H.; Keen, C. L. (2003). *Nature.* Nutrition: milk and absorption of dietary flavanols, 426, 787-788.

Shiina, Y.; Funabashi, N.; Lee, K.; Murayama, T.; Nakamura, K.; Wakatsuki, Y.; Daimon, M.; Komuro, I. (2009). *International Journal of Cardiology.* Acute effect of oral flavonoid-rich dark chocolate intake on coronary circulation, as compared with non-flavonoid white chocolate, by transthoracic Doppler echocardiography in healthy adults, 131, 424–429.

Spencer, J. P.; Schroeter, H.; Shenoy, B.; Srai, S. K.; Debnam E.; Rice-Evans, C. (2001). *Biochemical and Biophysical Research Communications.* Epicatechin is the primary bioavailable form of the procyanidin dimmers B2 and B5 after transfer across the small intestine, 285, 588-593.

Tabernero, M.; Serrano, J.; Saura-Calixto, F. (2006). *International Journal of Food Science & Technology.* The antioxidant capacity of cocoa products: contribution to the Spanish diet, 41, 28-32.

Tomaru, M.; Takano, H.; Osakabe, N.; Yasuda, A.; Inoue, K.; Yanagisawa, R.; Ohwatari, T.; Uematsu, H. (2007). *Nutrition.* Dietary supplementation with cacao liquor proanthocyanidins prevents elevation of blood glucose levels in diabetic obese mice, 23, 351-355.

United States Department of Agriculture. (USDA). *Nutrient Data Laboratory.* 2008 [2008 October 13th]. Available from: URL: http://www.nal.usda.gov/fnic/foodcomp/search/.

Vinson, J. A.; Proch, J.; Zubik, L. (1999). *Journal of Agricultural and. Food Chemistry.* Phenol antioxidant quantity and quality in foods: cocoa, dark chocolate, and milk chocolate, 47, 4821-4824.

Vinson, J. A.; Proch, J.; Bose, P.; Muchler, S.; Taffera, P.; Shuta, D.; Samman, N.; Agbor, G. A. (2006). *Journal of Agricultural and Food Chemistry.* Chocolate is a powerful *ex vivo* and *in vivo* antioxidant, an antiatherosclerotic agent in an animal model, and a significant contributor to antioxidants in the European and American Diets, 54, 8071-8076.

Wan, Y.; Vinson, J. A.; Etherton, T. D.; Proch, J.; Lazarus, S. A.; Kris-Etherton, P. M. (2001). *The American Journal of Clinical Nutrition.* Effects of cocoa powder and dark chocolate on LDL oxidative susceptibility and prostaglandin concentrations in humans, 74, 596–602.

Wiseman, S. A.; Balentine, D. A.; Frei, B. (1997). Critical Review Food Science Nutrition. *Antioxidants in tea.* 37, 705-718.

World Health Organization (WHO). *Diabetes: key facts*. 2017 [2017 November 10th]. Available from: URL: http://www.who.int/mediacentre/factsheets/fs312/en/.

Yuan, S.; Li, X.; Jin, Y.; Lu, J. (2017). *Nutrients*. Chocolate consumption and risk of coronary heart disease, stroke, and diabetes: A meta-analysis of prospective studies, 9, 688.

In: The Diversified Benefits of Cocoa … ISBN: 978-1-53613-258-8
Editor: Bonifacia Zayas Espinal

Chapter 3

CHOCOLATE AS A CARRIER FOR COCOA'S FUNCTIONAL INGREDIENTS

Maja Bulatović[1,*], Danica Zarić[2], Marica Rakin[1], Tanja Krunić[3], Ivana Lončarević[4] and Biljana Pajin[4]

[1]Faculty of Technology and Metallurgy, University of Belgrade, Belgrade, Serbia

[2]IHIS Techno experts d.o.o., Research Development Center, Belgrade, Serbia

[3]Innovation center Faculty of Technology and Metallurgy, University of Belgrade, Belgrade, Serbia

[4]Faculty of Technology, University of Novi Sad, Novi Sad, Serbia

ABSTRACT

The aim of this chapter was to introduce chocolate as a cocoa carrier of probiotics and bioactive whey protein hydrolysate. This research examined the influence of additional ingredients on the functional properties and rheology of the final product.

---

* Corresponding Author Email: mbulatovic@tmf.bg.ac.rs.

The chocolate prepared with *L. rhamnosus* HN001 strain at 40°C achieved the highest survivability, which was above 90% (viable cell count of about 8.1 log (CFU/g)) after 6 months of storage. This strain could be selected for high-quality probiotic chocolate production. The addition of probiotic *L. rhamnosus* HN001 significantly improves the rheological properties of probiotic chocolate. The addition of bioactive whey protein hydrolysate did not show significant changes in the total content of polyphenols or ABTS radical scavenging activity. However, DPPH radical scavenging activity showed an increase of 1.28 times during storage and about 1.46 in relation to the control sample and the sample supplemented with unhydrolyzed whey proteins. The bioactive whey protein hydrolysate affects chocolate rheology by increasing the thixotropic loop 8.02 times in relation to the control sample, and 1.57 times compared to the sample supplemented with unhydrolyzed whey proteins.

Based on the results, the addition of probiotics and bioactive whey protein hydrolysate powder in industrial conditions, which was done in the mixing tank (at 40°C) before the chocolate shaping phase, provides preservation of both functional ingredients as key components for this type of product. The addition of probiotics improves the rheological properties of chocolate, while the addition of whey protein hydrolysate leads to undesirable viscosity increase that can be avoided using a more efficient emulsifier.

Keywords: cocoa, chocolate, probiotics, antioxidants, whey

Introduction

Chocolate is a complex rheological system that consists of a high concentration (50–60% by vol.) of suspended solid particles, which are comprised mostly of sugar crystals, cocoa and milk solids and these are dispersed in a continuous fat matrix made of cocoa butter and milk fats (Beckett, 2008). Prominent substances such as cacao based polyphenolic compounds as original healthy components inherent in cocoa are usually lost during processing (Konar et al., 2016). Therefore, the addition of functional ingredients could be a promising way to compensate this loss of health potential. One of the opportunities is the addition of probiotics and bioactive peptides as ingredients that can significantly affect functional properties of chocolate. There are nearly 20 known bacterial species that beneficially affect the balance of more than 400 different microorganisms that naturally inhabit the human digestive system (Gupta and Garg, 2009). Many of these have already been successfully included in fermented dairy products, but their use in chocolate products is still a challenge.

Due to the fact that the necessary recommended dose of probiotics needs to be at least 10^6-10^7 CFU per gram (Hernandez-Mendoza et al., 2007), the viability of probiotics during production and storage of chocolate is a special issue that needs to be dealt with considering the sensitivity of these microorganisms to temperature and aerobic conditions. As these products are often exposed to oxygen, researchers have so far studied survivability of probiotic strains during storage at various temperatures. Possemiers et al., (2010) have reported the synergistic system of bacteria and chocolate as a successful combination for probiotic delivery. They have showed that coating the probiotics with chocolate is an excellent solution to protect them from stressful environmental conditions and for optimal delivery. In addition, based on the findings reported by Aragon-Alegro et al., (2007), during a short storage period of 28 days, as well as the influence of prebiotic inulin, an increase in the viable cell count of the *Lactobacillus paracasei* strain incorporated in chocolate mousse could be achieved. The incorporation of probiotic cells (encapsulated by spray-coating technology) in chocolate, revealed that probiotic viability in the small intestine was three times higher when incorporated in chocolate than in dairy products (Maillard & Landuyt, 2008). Also, the lipid fraction of cocoa butter, as the main component of chocolate, was shown to be protective for bifidobacteria (Burgain et al., 2011; Lahtinen et al., 2007). Yonejima et al., (2015) have noted that probiotics incorporated in chocolate were found to be much more resistant to the conditions of the gastrointestinal environment than probiotic powder or those delivered through beverages or yoghurt.

Apart from probiotics, whey protein hydrolysates were recently recognized as a food ingredient with important nutritional and functional properties gaining acceptance as a valuable functional food ingredient. Hydrolysis of whey proteins is known to release bioactive peptides that can exhibit a number of physiological properties and enable them to demonstrate antioxidative, antihypertensive, antimicrobial, opioid and antithrombotic activity (Park and Nam, 2015). However, the number of researches related to the production of functional chocolate and its placement in commercial flows is quite small, which could not be explained by the inappropriateness of the chocolate matrix (Konar et al., 2016). A possible answer to this scarcity may be attributed to the influence of incorporated functional ingredients on the rheology of chocolate as a parameter that is equally important both for the production process and for the sensory quality of the final product. According to literature reports (Laličić-Petronijević et al., 2015; Żyżelewicz et al., 2015), incorporation of additional ingredients usually increases the rheological parameters of chocolate and negatively influences its flow properties. In addition, there could be a significant correlation between interactions of polyphenols and proteins, but the interference of proteins with polyphenols remains inconclusive, considering the possibility that proteins could reduce the activity of polyphenols (Serafini & Crozier, 2003; Tabernero et al., 2006).

However, due to the grinding stage, friction and high temperatures can affect the viability of probiotic bacteria and the activity of bioactive peptides, they cannot be added during composing of the milk chocolate - at the beginning of the industrial process (Zarić et al., 2016). The achievement of satisfactory rheology without damaging the probiotics and peptides, demands their incorporation at the process stage that allows full homogenization of the chocolate mass. Therefore, the addition of functional ingredients in industrial conditions should be done in the mixing tank (at lower temperatures) before the phase of chocolate shaping.

The aim of this chapter was to introduce chocolate as a cocoa carrier of probiotics and bioactive whey protein hydrolysate. This research examined the influence of the additional ingredients on the functional properties and rheology of the final product.

Materials and Methods

Milk Chocolate Mass

Raw materials used in the production of chocolate mass were: sugar 41.5% (Crvenka AD, Serbia), dairy milk powder 25.4% (fat 25%, protein 28%) (Imlek, Serbia), cocoa butter 18.9% (Theobroma, The Netherlands), cocoa mass 10.4% (Cargill, Ghana), skimmed milk powder (fat 1.7%, protein 35%) (Imlek, Serbia), lecithin 0.5% (Soyaprotein AD, Serbia), and flavoring 0.06% (Etol, Slovenia). The milk chocolate mass had 1.1 ± 0.06% moisture.

Probiotic Microorganisms

Concentrated and freeze-dried probiotic strains *Lactobacillus acidophilus* NCFM (Howaru® Dophilus), *Bifidobacterium lactis* HN019 (Howaru® Bifido) and *Lactobacillus rhamnosus* HN001 (Howaru® Rhamnosus) were obtained from Danisco, (Madison, WI, USA). The lyophilized strains were inoculated in the proportions 2.5 DCU/kg, 2.5 DCU/kg and 5.0 DCU/kg (respectively), according to the manufacturer recommendations, to obtain a functional level of probiotics (at least 10^6-10^7 CFU/g).

Bioactive Whey Protein Hydrolysate

Bioactive whey protein hydrolysate (BWPH) was produced according to the method of enzymatic hydrolysis described by Embiriekah et al., (2017) using commercial enzyme trypsin (porcine pancreas, EC 3.4.21.4) purchased from Sigma-Aldrich Chemie GmbH (Steinheim, Germany). After the hydrolysis, the produced hydrolysate was spray dried using a laboratory scale Mini spray dryer B-290 (BUCHI, Labortechnik AG,

Switzerland), in conjunction with a peristaltic pump, under the following conditions: inlet temperature at 120°C, corresponding to a reading of outlet temperature 68-70°C. The batch of the powdered sample was stored in a desiccator (at room temperature) before use. Bioactive whey protein hydrolysate (BWPH) powder was added to the chocolate mass in a concentration of 6.0% (w/w).

Production of Milk Chocolate Mass

The chocolate mass was produced in the laboratory ball mill with a homogenizer (capacity 5 kg). The raw materials were measured and simultaneously dosed into the homogenizer (except 10% of cocoa butter, which is dosed 10 min before removing the mass from the ball mill) and mixed for 20 min at a temperature of 50°C and mixer rotation speed of 50 rpm. The homogenous mass was then transferred into the ball mill (ball diameter 9.1 mm; ball mass 30 kg; mixer rotation speed 50 rpm; mill inner diameter 0.250 m; height 0.31 m.; the volume of space provided for the balls and 5 kg of chocolate mass was 0.0152 m^3). Applied refining time in the mill was 90 min.

In order to avoid exposure of probiotic bacteria and bioactive whey protein hydrolysate to the high temperatures and harmful effects of mechanical shear during milling in the ball mill, ingredients were introduced into the chocolate mass after milling, i.e., before the pre-crystallization and molding phases, when the temperature was reduced. The probiotics and BWPH were added at a temperature of 40°C, in the above prescribed concentrations. The temperature of the chocolate mass was sustained in the mixer in a water bath for 15 min, by mixing with a minimal number of rotations of the blender.

After adding the ingredients the pre-crystallization process was carried out. Pre-crystallization of the chocolate mass was performed in the laboratory pre-crystallizer, a modified Brabender farinograph (Pajin, 2009; Jovanović and Pajin, 2004). The process of pre-crystallization was controlled indirectly by changes of the mass resistance during mixing, which was registered on a force/time diagram - the thermorheogram. Torque value is a criterion for the viscous behavior of the chocolate mass and is dependent on the crystallization extent of the mass in question. The pre-crystallization temperature of 28°C was applied. The pre-crystallized mass was then molded, cooled and removed from the forms. The final products were packed in aluminium foil and in marked white paper/cardboard, and then stored at 20 ± 2°C. The chemical composition of the final chocolate tablets was determined using the standard AOACC (2000) method. The chemical composition of the final probiotic chocolate tablets was as follows: 9.41 ± 0.10% proteins, 30.85 ± 0.07% fats, 53.82 ± 0.17% sugars and 1.1 ± 0.05% moisture. The chemical composition of the final functional chocolate tablets supplemented with bioactive whey protein hydrolysate was as follows: 14.52 ± 0.10% proteins, 29.1 ± 0.07% fats, 50.77 ± 0.17% sugars and 1.1 ± 0.05% moisture.

Viability of Probiotic Bacteria

The amount of 1 g of investigated chocolate samples was dissolved in 9 mL of sodium chloride solution (0.85%, w/v) at 40°C, and mixed uniformly. Subsequent serial dilutions were prepared and the viable cell count was determined using pour plate technique on MRS agar (Vrbaški & Markov, 1993). The plates were incubated at 37°C for 48 h in anaerobic conditions. Probiotic bacteria were enumerated as colony forming units per gram of chocolate and expressed as log (CFU/g). Viability tests were performed after 0, 30, 60, 90, 120, 150 and 180 days of storage at 20 ± 2°C.

Antioxidant Activity of the Produced Chocolate Mass

The *total polyphenol content* in the chocolate samples was determined spectrophotometrically by the Folin - Ciocalteau method described by Đorđević et al., (2010) with slight modifications. Briefly, a chocolate sample (1g) was dissolved in 9 mL of distilled water. The prepared solution (100 µl) of the investigated sample was shaken for 60s on the Vortex mixer with 500 µL of Foline-Ciocalteu reagent and 6 mL of distilled water. After that, 2 mL of 15% Na_2CO_3 was added and the mixture was shaken once again for 30s. Finally, the solution was brought up to 10 mL by adding distilled water. After 2 h, the absorbance was read on the UV/visible spectrophotometer (Ultrospec 3300 pro, Amersham Bioscience, Uppsala, Sweden) at 750 nm (25°C) using glass cuvettes against a blank solution (100 µL of distilled water instead of test sample). The results were calculated using a gallic acid calibration curve and expressed as gallic acid equivalents (GAE) per 100 g of sample.

The *ABTS radical scavenging activity* of chocolate samples was tested according to the method described by Re et al., (1999) with slight modifications. This method is based on the ability of antioxidant molecules to quench the stable bisradical cation, ABTS•+, a blue/green chromophore with characteristic absorption at 734 nm. The ABTS radical cation (ABTS•+) was produced in the reaction between ABTS (7 mM) and potassium persulfate (2.45 mM final concentration). The reaction mixture was kept in the dark at room temperature for at least 12 h before use. Further, the ABTS•+ solution was diluted with 5 mM phosphate-buffered saline (PBS, pH 7.4), until the absorbance of 700 (±0.05) was reached. The chocolate samples (1g) were dissolved in 9 mL of distilled water. The prepared samples (10 µL) were mixed with 990 µL of diluted ABTS•+ solution and absorbance was measured at 734 nm after 5 min. The ABTS radical scavenging activity (%) was calculated using the following Eq. 1:

$$\text{ABTS radical scavenging activity (\%)} = \frac{Ac - As}{Ac} \times 100 \qquad (1)$$

where: A_C represents the absorbance of the control solution and A_S is the absorbance of the sample solution.

The *DPPH radical scavenging activity* of chocolate samples was tested according to the method described by Balakrishnan and Agrawal (2014) with slight modifications. Briefly, a chocolate sample (2g) was dissolved 8 mL of distilled water. The dissolved sample was mixed with methanol in a ratio of 1:3 and centrifuged at 10000 x g for 10 minutes. The obtained supernatant (100 μL) was mixed with 900 μL of DPPH solution (0.1 mM solution of 2,2-diphenyl-1-picylhydrazine in 95.0% methanol). The absorbance of the solution was measured after 30 min at 517 nm against methanol as a blank solution. The control sample was prepared using 100 μL of methanol instead of the sample. The DPPH radical scavenging activity (%) was calculated using the following Eq. 2:

$$\text{DPPH radical scavenging activity (\%)} = \frac{\text{Ac - As}}{\text{Ac}} \times 100 \qquad (2)$$

where: A_C represents the absorbance of the control solution and A_S is the absorbance of the sample solution.

Antioxidant activity tests were performed immediately after production (0 day) and after 180 days of storage at 20 ± 2°C.

Rheological Properties of the Produced Chocolate Mass

Rheological properties were determined in the rotation viscometer RheoStress (600 HP, Haake, Germany), according to the IOCCC (2000) method, at a temperature of 40 ± 0.1°C. Flow curves were determined using the method of the hysteresis loop within the shear rate interval of 1-60 1/s. Shear rate was increased from 1 to 60 s^{-1} during the period of 240 s, then maintained at the maximum speed of 60 s^{-1} for 60 s, and the decreasing of shear rate from 60 to 1 s^{-1} also lasted 240 s.

Statistical Analysis

All experiments were performed in triplicate. Mean values were analyzed using one-way ANOVA. The Tukey post hoc test was performed for means comparison (OriginPro 8, Origin Lab Co., Northampton, USA). Differences were considered as significant at $P < 0.05$.

Results and Discussion

Viability of Probiotic Bacteria in Probiotic Chocolate Mass

Probiotic microorganisms are the most sensitive factor in the probiotic chocolate production process. Their viability is the crucial parameter related to achieving and maintaining the functional properties of probiotic chocolate. The advantage of the chocolate matrix for carrying probiotics could be associated with the high polyphenolic content of cacao. In addition to exhibiting antioxidant activity, chocolate might serve as a better probiotic carrier than dairy products for intestinal delivery (Erdem et al., 2014). Changes in viable cell count of probiotic bacteria in chocolate samples, during storage at 20 ± 2°C, are presented in Figure 1.

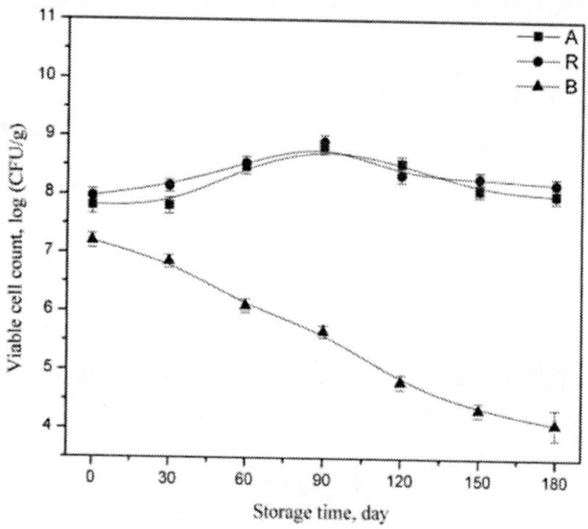

Figure 1. Changes in viable cell count of probiotic bacteria in the chocolate samples during storage at 20 ± 2°C. Symbols: ■ *L. acidophilus* NCFM (A), ● *L. rhamnosus* HN001 (R) and ▲ *B. lactis* HN019 (B).

As indicated in Figure 1, initial viable cell count of probiotic bacteria ranged from 7.0 to 7.97 log (CFU/g). It was at a satisfactory level ($\geq 10^6$ CFU/g) recommended (Hernandez-Mendoza et al., 2007) for probiotic products in all the chocolate samples. Gradual decrease of viable cell count was observed in the chocolate sample cultured with strain *B. lactis* HN019 (B), during the whole storage period. This behavior could be explained by the fact that this strain is highly sensitive to oxygen exposure. The mixing phase (which is characterized by the presence of high oxygen levels) applied during chocolate production could be the reason of its poor viability. It could be concluded that strain *B. lactis* HN019 is capable to survive in chocolate, at a desirable level, for no longer than 60 days of storage at 20 ± 2°C. Observed results are slightly lower than those reported in literature (Laličić-Petronijević et al., 2015) concerning the viability of the *B. lactis* HN019 strain, probably due to the lower initial cell count. Based on these findings, it could be concluded that initial cell count is a crucial parameter for achieving functionality of probiotic chocolate cultured with the highly sensitive *B. lactis* HN019 strain.

On the other hand, viable cell count in samples cultured with strains *L. rhamnosus* (R) and *L. acidophilus* (A) remained at a satisfactory level, even greater than recommended ($\geq 10^6$ CFU/g), during the whole storage period. Viable cell count gradually increases during the 90 days of storage, reaching the maximal viable cell count of about 8.85 log CFU/g in samples A and R. The obtained results are consistent with those reported in literature (Laličić-Petronijević et al., 2015) concerning the viability of *L. acidophilus* NCFM, where the viable cell count remains at the same level of about 8.6 log CFU/g during the whole storage period. It is interesting to note that viable cell count reported in literature (Laličić-Petronijević et al., 2015) gradually decreases during the 90 days in contrast to the result obtained in the present study. It could be explained by the fact that initial viable cell count reported in our study was not at its maximal level, which allowed subsequent growth of *L. rhamnosus* and *L. acidophilus* strains.

Among the samples, after 90 days of storage, samples A and R had a significantly ($P < 0.05$) greater viable cell count than sample B. The same behavior was observed in these samples after 120 days of storage. Also, comparing results related to cell growth in production treatment carried out at a temperature of 30-32°C reported in literature (Laličić-Petronijević et al., 2015), it could be said that the temperature of 40°C had significantly positive influence on viability of *L. rhamnosus* and *L. acidophilus* strains, by improving their activity. Based on the observed results, it could be concluded that chocolate cultured with *L. rhamnosus* and *L. acidophilus* strains at 40°C, exhibits high functional quality, and these strains express a great potential for use in production of probiotic chocolate.

Rheological Properties of Probiotic Chocolate Mass

Rheological parameters such as Casson viscosity and Casson yield value of chocolate mass are of key importance for the manufacturing technology. In industrial processes, these quantities should be as low as possible to decrease resistance during unit processes such as mixing or pumping (Beckett, 2008). Table 1 shows the rheological parameters of probiotic chocolate samples determined by static measurements.

Table 1. Rheological parameters of the chocolate samples produced without probiotics (C), with *Lactobacillus acidophilus* NCFM (A), *Bifidobacterium lactis* HN019 (B) and *Lactobacillus rhamnosus* HN001 (R) determined by static measurements

Sample	Thixotropic curve area, Pa/s	Casson yield stress, Pa	Casson viscosity, Pa·s
C	2776^{c}	23.62^{a}	3.16^{e}
A	1751^{d}	23.71^{a}	2.49^{b}
B	1289^{e}	21.08^{e}	2.35^{bc}
R	1089^{f}	19.93^{f}	2.25^{c}

Values in the same column with the same superscript are not statistically different ($P > 0.05$).

In comparison to the chocolate sample without probiotics (C), chocolate samples B and R have lower viscosity and yield stress. The range of viscosity in the chocolate samples varied from 2.25 to 2.49 Pa·s while the yield stress from 19.93-23.71 Pa. Based on the results, it could be assumed that the probiotic strain *L. rhamnosus* causes the least negative effect on the rheological properties of the produced probiotic chocolate. However, viscosity of samples with *L. acidophilus* and *B. lactis* inoculated at 40°C was lower than the viscosity of samples inoculated at 30-32°C reported in literature (Laličić-Petronijević et al., 2015). Based on these findings, it could be said that the addition of probiotic strains in conjunction with a temperature of 40°C positively affects rheological parameters of probiotic chocolate and leads to the improvement of flow properties.

Antioxidant Activity of Bioactive Chocolate Mass

Not only does treatment with antioxidants offer a valuable preservation technique that can prolong the shelf life of food, also the application of natural antioxidants as dietary supplements attracts great attention based on their therapeutic properties. The antioxidant activity of chocolate samples produced by the addition of unhydrolyzed (WPC) and hydrolyzed (BWPH) whey protein powder is presented in Table 2.

As shown in Table 2 the addition of unhydrolyzed whey proteins (WPC) and bioactive whey protein hydrolysate (BWPH) increases the content of total polyphenols 1.15 times compared to the control sample (C). In addition, inhibition of ABTS radicals increases 1.37 times with the addition of BWPH, while WPC supplementation did not significantly affect the ABTS radical scavenging activity. The DPPH test showed an increase in inhibition of DPPH radical by addition of BWPH (1.17 times), while WPC supplementation did not show a significant influence on this parameter. After six months of storage at 20 ± 2°C, both samples (WPC and BWPH) achieved an increase in antioxidant properties. The total content of polyphenols and ABTS radical scavenging activity did not show statistically significant ($P < 0.05$) changes, while DPPH radical scavenging activity in the BWPH sample showed an increase of 1.28 times compared to the initial value, and about 1.46 relative to C and WPC samples.

Table 2. Antioxidant activity indicators of chocolate samples produced without supplements (C) and with the addition of unhydrolyzed (WPC) and hydrolyzed (BWPH) whey protein powders, immediately after production (0 day) and after 180 days of storage at 20 ± 2°C

Sample	Time, day	Total polyphenol content, mg GAE/100g	ABTS, %	DPPH, %
C	0	877.1 ± 56.0	48.46 ± 2.11	48.07 ± 2.80
	180	932.0 ± 44.0	50.11 ± 2.52	49.21 ± 2.52
WPC	0	1012.1 ± 67.3	51.01 ± 2.24	46.43 ± 4.17
	180	981.5 ± 35.0	53.30 ± 1.99	48.25 ± 3.50
BWPH	0	1007.8 ± 96.8	66.30 ± 1.99	56.34 ± 3.20
	180	1088.2 ± 48.0	69.90 ± 1.52	72.04 ± 4.10

Table 3. Rheological parameters of the chocolate samples produced without supplements (C) and with the addition of unhydrolyzed (WPC) and hydrolyzed (BWPH) whey protein powders

Sample	Thixotropic curve area, Pa/s	Casson yield stress, Pa	Casson viscosity, Pa·s
C	1073^{c}	12.81^{c}	3.53^{c}
WPC	5499^{b}	20.08^{b}	9.17^{b}
BWPH	8608^{a}	23.40^{a}	11.81^{a}

Values in the same column with the same superscript are not statistically different ($P > 0.05$).

Rheological Properties of Bioactive Chocolate Mass

The chocolate mass is a non-Newtonian fluid that shows plastic flow and has a characteristic yield stress representing the force that must be applied to make the chocolate mass flow. Controlling the viscosity of the chocolate mass is necessary since it directly affects the quality and the price of the final product (Afoakwa et al., 2008). Table 3 shows the rheological parameters of functional chocolate samples determined by static measurements.

The results presented in Table 3 reveal a significant difference in the samples rheology, depending on the supplement type. The most homogeneous structure, with the lowest values of Casson yield stress, Casson viscosity and thixotropic curve area, has the chocolate sample without supplements (C). The sample supplemented with bioactive whey protein hydrolysate (BWPH) has greater deviations from that of the sample with unhydrolyzed whey proteins (WPC). The difference in the thixotropic curve areas of the three samples is clearly visible in Table 3. The size of the thixotropic loop depends on the ingredients of the chocolate mass, the fat content, processing temperature, and strength of the applied force for mixing (Afoakwa et al., 2008). All the chocolate masses examined in this chapter have the same initial composition and were identical in treatment. The difference in rheological properties is solely due to the presence of various protein supplements in the samples. The addition of WPC and BWPH, rich in peptides, influences the packing of particles. The solid particles did not occupy the proper place in the crystal lattice, which led to different intermolecular connections and the growth of Casson yield stress and viscosity. The addition of BWPH has a significant effect on the samples rheology and increases the thixotropic curve area 8.02 times in relation to the control sample (C), and 1.57 times compared to the sample supplemented with WPC. Supplementation with WPC increases the thixotropic curve area 5.12 times in relation to the control sample (C). With the addition of BWPH viscosity of the chocolate increases 3.37 times, while the addition of WPC increases viscosity 2.60 times in relation to the control sample (C). It is important to note that the addition of hydrolyzed and unhydrolyzed whey proteins affects viscosity more than Casson yield stress. The increase in viscosity can be reduced by a selection of emulsifiers with better emulsifying properties. The combination of emulsifiers most often used for the reduction of viscosity is the mixture composed of 0.2% PGPR (Polyglycerol polyricinoleate) and 0.3% lecithin (Pajin et al., 2011).

CONCLUSION

Based on the findings reported in this chapter, it can be concluded that chocolate mass can be successfully enriched with probiotic strains *Lactobacillus acidophilus* NCFM, *Lactobacillus rhamnosus* HN001 and *Bifidobacterium lactis* HN019. The chocolate prepared with *L. rhamnosus* HN001 strain at 40°C achieved highest survivability and this strain can be selected for high quality probiotic chocolate production. The addition of probiotic *L. rhamnosus* HN001 significantly improves the rheological properties of probiotic chocolate. Due to the high survivability of the *L. rhamnosus* HN001 strain after 6 months of storage, which was above 90% (viable cell count of about 8.1 log (CFU/g)), the produced chocolate can be considered as an excellent cocoa carrier of probiotics. The addition of bioactive whey protein hydrolysate did not show significant changes in the total content of polyphenols and ABTS radical scavenging activity, while DPPH radical scavenging activity showed an increase of about 1.46 times in relation to the control sample and the sample supplemented with unhydrolyzed whey proteins. During the 6 months of storage, DPPH radical scavenging activity of the chocolate sample supplemented with bioactive whey protein hydrolysate increases about 1.28 times. The bioactive whey protein hydrolysate affects the chocolate rheology by increasing the thixotropic loop 8.02 times in relation to the control sample, and 1.57 times compared to the sample supplemented with unhydrolyzed whey proteins.

Based on the results, the addition of the probiotics and bioactive whey protein hydrolysate powder in industrial conditions, which was done in the mixing tank (at 40°C) before the phase of chocolate shaping, provides preservation of both functional ingredients as key components of this type of product. The addition of probiotics improves the rheological properties of chocolate, while the addition of whey protein hydrolysate leads to undesirable viscosity increase that can be avoided using a more efficient emulsifier.

ACKNOWLEDGMENTS

This work was supported by the Ministry of Education, Science and Technological Development (Project numbers: TR 31017 and TR 31014).

REFERENCES

Afoakwa, E. O., Paterson, A., & Fowler, M. (2008). Effects of particle size distribution and composition on rheological properties of dark chocolate. *European Food Research and Technology, 226,* 1259-1268.

AOAC (2000). *Official methods of analysis of the association of official analytical chemists international. Vols. I and II.* 17th ed. Arlington: Association of Official Analytical Chemists Inc.

Aragon-Alegro, L. C., Alegro, J. H. A., Cardarelli, H. R., Chiu, M. C., & Saad, S. M. I. (2007). Potentially probiotic and synbiotic chocolate mousse. *LWT-Food Science and Technology, 40,* 669-675.

Balakrishnan, G., & Agrawal, R. (2014). Antioxidant activity and fatty acid profile of fermented milk prepared by Pediococcus pentosaceus. *Journal of Food Science and Technology, 51,* 4138-4142.

Beckett, S. T. (2008). *Science of Chocolate* (2nd ed.). Cambridge, RSC Publishing.

Burgain, J., Gaiani, C., Linder, M., & Scher, J. (2011). Encapsulation of probiotic living cells: From laboratory scale to industrial applications. *Journal of Food Engineering, 104,* 467-483.

Đorđević, T. M., Šiler-Marinković, S. S., & Dimitrijević-Branković, S. I. (2010). Effect of fermentation on antioxidant properties of some cereals and pseudo cereals. *Food Chemistry, 119,* 957-963.

Embiriekah, S., Bulatović, M., Borić, M., Zarić, D., & Rakin, M. Antioxidant activity, functional properties and bioaccessibility of whey protein hydrolysates. *International Journal of Dairy Technology,* doi:10.1111/1471-0307.12428.

Erdem, Ö., Gültekin-Özgüven, M., Berktaş, I., Erşan, S., Tuna, H. E., Karadağ, A., Özçelik, B., Güneş, G. & Cutting, S. M. (2014). Development of a novel synbiotic dark chocolate enriched with *Bacillus indicus* HU36, maltodextrin and lemon fiber: Optimization by response surface methodology. *LWT-Food Science and Technology, 56,* 187-193.

Gupta, V., & Garg, R. (2009). Probiotics. *Indian Journal of Medical Microbiology, 27,* 202-209.

Hernandez-Mendoza, A., Robles, V. J., Angulo, J. O., De La Cruz, J., & Garcia, H. S. (2007). Preparation of a whey-based probiotic product with Lactobacillus reuteri and *Bifidobacterium bifidum. Food Technology and Biotechnology, 45,* 27-31.

IOCCC (2000). Viscosity of cocoa and chocolate products. Analytical Method 46-2000., *International Office of Cocoa, Chocolate and Confectionary*, Geneva.

Jovanović, O., & Pajin, B. (2004). Influence of lactic acid ester on chocolate quality. *Trends in food science and technology*, *15*, 128-136.

Konar, N., Toker, O. S., Oba, S., & Sagdic, O. (2016). Improving functionality of chocolate: A review on probiotic, prebiotic, and/or synbiotic characteristics. *Trends in Food Science and Technology, 49,* 35-44.

Lahtinen, S. J., Ouwehand, A. C., Salminen, S. J., Forssell, P., & Myllärinen, P. (2007). Effect of starch- and lipid-based encapsulation on the culturability of two *Bifidobacterium longum* strains. *Letters in Applied Microbiology, 44,* 500-505.

Laličić-Petronijević, J., Popov-Raljić, J., Obradović, D., Radulović, Z., Paunović, D., Petrušić, M., & Pezo, L. (2015). Viability of probiotic strains *Lactobacillus acidophilus* NCFM® and *Bifidobacterium lactis* HN019 and their impact on sensory and rheological properties of milk and dark chocolates during storage for 180 days. *Journal of Functional Foods, 15,* 541-550.

Maillard, M., & Landuyt, A. (2008). Chocolate: an ideal carrier for probiotics. *Agro Food Industry Hi-Tech, 19,* 13-15.

Pajin, B. (2009). *Analysis for confectionery products* (1st ed.). Novi Sad, Faculty of Technology.

Pajin, B., Zarić, D., Dokić, L., Šereš, Z., Šoronja-Simović, D., Omorjan, R., & Lončarević, I. (2011). Influence of emulsifiers on the optimization of processing parameters of refining milk chocolate in the ball mill. *Acta Periodica Technologica, 42,* 101-110.

Park, Y. W., & Nam, M. S. (2015). Bioactive peptides in milk and dairy products: a review. *Korean Journal for Food Science of Animal Resources, 35,* 831.

Possemiers, S., Marzorati, M., Verstraete, W., & Van de Wiele, T. (2010). Bacteria and chocolate: a successful combination for probiotic delivery. *International Journal of Food Microbiology, 141,* 97-103.

Re, R., Pellegrini, N., Proteggente, A., Pannala, A., Yang, M., & Rice-Evans, C. (1999). Antioxidant activity applying an improved ABTS radical cation decolorization assay. *Free Radical Biology and Medicine, 26,* 1231-1237.

Serafini, M., & Crozier, A. (2003). Nutrition: milk and absorption of dietary flavanols. *Nature*, *426*, 788-788.

Tabernero, M., Serrano, J., & Saura-Calixto, F. (2006). The antioxidant capacity of cocoa products: contribution to the Spanish diet. *International Journal of Food Science and Technology*, *41*, 28-32.

Vrbaški, L., & Markov, S. (1993). *Practicum of microbiology*, Novi Sad, Prometej.

Yonejima, Y., Hisa, K., Kawaguchi, M., Ashitani, H., Koyama, T., Usamikrank, Y., & Ogawa, J. (2015). Lactic acid bacteria-containing chocolate as a practical probiotic product with increased acid tolerance. *Biocatalysis and Agricultural Biotechnology*, 4, 773-777.

Zarić, D., Bulatović, M., Rakin, M., Krunić, T., Lončarević, I., Pajin, B. (2016). Functional, rheological and sensory properties of probiotic milk chocolate produced in a ball mill, *RSC Advances*, 6, 13934–13941.

Żyżelewicz, D., Nebesny, E., Motyl, I., & Libudzisz, Z. (2010). Effect of milk chocolate supplementation with lyophilised *Lactobacillus* cells on its attributes. *Czech Journal of Food Sciences*, 28, 392-406.

Biographical Sketch

Name: Maja Bulatović

Affiliation: Faculty of Technology and Metallurgy, University of Belgrade, Belgrade, Serbia.

Education:
Degree: MSc
Institution: Faculty of Technology and Metallurgy, University of Belgrade
Field: Biochemical Engineering and Biotechnology
Year: 2007.

Degree: PhD
Institution: Faculty of Technology and Metallurgy, University of Belgrade
Field: Technological engineering - Biotechnology
Year: 2015.

Business Address: Faculty of Technology and Metallurgy, Karnegijeva 4, 11000 Belgrade, Serbia

Research and Professional Experience:
Upon completion of studies, started to work in the paint and varnish factory "Jugohem" d.o.o, as a quality manager in research and development of new products.

Since February 1st 2011, employed as a researcher at the Faculty of Technology and Metallurgy, at the Department of Biochemical Engineering and Biotechnology. Currently involved in the project of Technology Development entitled "*Production of lactic acid and probiotics in waste products of food and agro-industry,*" financed by the Ministry of Education, Science and Technological Development of the Republic of Serbia.

In the period 2012-2013 participated in the realization of the Innovation Project entitled "*Fermented whey-based beverages as a new functional dairy products,*" financed by the Ministry of Education, Science and Technological Development of the Republic of Serbia, resulting in the Technical Solution entitled "*Production of functional fermented beverage made from whey and milk*" which was adopted by AD Imlek - Belgrade as end user.

In the period 2014-2015 participated in the realization of the Innovation Project entitled "*Production and application of bioactive proteins and whey and milk peptides,*" financed by the Ministry of Education, Science and Technological Development of the Republic of Serbia, resulting in the Technical Solution entitled "*Improvement of functional characteristics of fermented whey and milk beverage by the addition of bioactive peptides*" which was adopted by foreign milk industry AD Bimlek - Macedonia as the end user.

My long-term research interests focus on the field of fermentation, especially lactic acid, which is aimed at the production of beverages from the category of functional foods as well as valorization of whey and other by-products and waste products of the food industry. In addition, special interest is the development of functional foods as new advanced food products as well as the accompanying production processes. During the doctoral program at the University of Belgrade Faculty of Technology and Metallurgy, I have collaborated with colleagues at University of Novi Sad Faculty of Technology and IHIS Techno-experts Research development center on qualitative research designed to develop chocolate as most popular confectionery product. In the period 2014-2015 with colleagues from the Institute for Hygiene with Medical Ecology, Faculty of Medicine,

University of Belgrade, I was involved in the realization of the pilot study entitled "*Application of whey in overweight people and monitoring of effects on appetite/weight control and weight regulation*" As an associate I am engaged in carrying out practical teaching in several subject areas in the field of biotechnology. During the research work actively participated in the work with students resulting with more than 30 master papers.

Professional Appointments:

2016- Research Associate, Faculty of Technology and Metallurgy, University of Belgrade

2012-2016 Research Assistant, Faculty of Technology and Metallurgy, University of Belgrade

2011-2012 Research Trainee, Faculty of Technology and Metallurgy, University of Belgrade

2007-2010 Quality manager, paint and varnish factory "Jugohem" d.o.o

Honors:

2013 Award for creating a new eco-innovation fermented beverage named "Active Drink" at the first competition of student teams in the creation of eco-innovative food products - ECOTROPHELIA EUROPE - SERBIA, organized by the Association of Food Technologists of Serbia.

Publications from the Last 3 Years:

Slavica Arsić, Maja Bulatović, Danica Zarić, Gordana Kokeza, Jonel Subić, Marica Rakin (2017). Functional fermented whey carrot beverage - qualitative, nutritive and techno-economic analysis. *Romanian Biotechnological Letters*, (ISSN 1224-5984), IF (2015) = 0.478.

Salem Embiriekah, Maja Bulatović, Milka Borić, Danica Zarić and Marica Rakin (2017) Antioxidant activity, functional properties and bioaccessibility of whey protein hydrolysates, *International Journal of Dairy Technology*, DOI: 10.1111/1471-0307.12428 (ISSN 1471-0307)

Tanja Krunić, Marica Rakin, Maja Bulatović and Danica Zarić (2018.) The contribution of bioactive peptides of whey to quality of food products, *Food Processing for Increased Quality and Consumption*, 1st Edition, Eds. A. Grumezescu and A. M. Holban, Academic Press-Elsevier, February 28, 2018, pp. (ISBN 978-0-1281-1447-6).

Rakin Marica B., Bulatović Maja Lj., Zarić Danica B., Stamenković-Đoković Marijana M., Krunić Tanja Ž., Borić Milka M., Vukašinović-Sekulić Maja S. (2016). Quality of fermented whey beverage with milk, *Chemical Industry*, 70 (1) 91-98 (ISSN 0367-598) (IF (2014) = 0.364, Engineering, Chemical, 121/135).

Krunić Tanja Ž., Obradović Nataša S., Bulatović Maja Lj., Vukašinović-Sekulić Maja S., Trifković Kata T., Rakin Marica B. (2016). Impact of carrier material on fermentative activity of encapsulated yoghurt culture in whey based substrate, *Chemical Industry*, DOI: 10.2298/HEMIND150717016K (ISSN 0367-598) (IF (2014) = 0.364, Engineering, Chemical, 121/135).

Maja Lj. Bulatović, Danica B. Zarić, Tanja Krunić, Milka Borić, Marica B. Rakin (2016). Production of the milk chocolate with probiotic in a ball mill. In: *Chocolate: Production, Consumption and Health Benefits*, Emily Baker (Ed.), Nova Science Publishers, New York, 77-103. (ISBN: 978-1-53610-447-9).

Danica B. Zarić, Maja Lj. Bulatović, Marica B. Rakin, Tanja Ž. Krunić, Ivana S. Lončarević, Biljana S. Pajin (2016). Functional, rheological and sensory properties of probiotic milk chocolate produced in a ball mill. *RSC Advances*, 6 (17), 13934 - 13941. (ISSN 2046-2069) (IF (2014) = 3.840, Chemistry, Multidisciplinary, 33/157).

Maja Vukašinović Sekulić, Marica Rakin and Maja Bulatović (2015). Fermented vegetable juices and health attributes, *Handbook of Vegetable Preservation and Processing*, 2nd Edition, Eds. Y. H. Hui and E. Özgül Evranuz, CRC Press, August 15, 2015, pp. 703-724 (ISBN 978-1-4822-1228-0).

Nataša S. Obradović, Tanja Ž. Krunić, Kata T. Trifković, Maja Lj. Bulatović, Marko P. Rakin, Marica B. Rakin, Branko M. Bugarski (2015). Influence of chitosan coating on mechanical stability of biopolymer carriers with probiotic starter culture in fermented whey beverages, *International Journal of Polymer Science*, DOI:10.1155/2015/732858. (ISSN 1687-9422) (IF (2015) = 1.746, Polymer Science, 51/82).

Tanja Ž. Krunić, Maja Lj. Bulatović, Nataša S. Obradović, Maja S. Vukašinović-Sekulić and Marica B. Rakin (2015). Effect of immobilization materials on viability and fermentation activity of dairy starter culture in whey based substrate, *Journal of the Science of Food and Agriculture*, DOI: 10.1002/jsfa.7278. (ISSN 0022-5142) (IF (2014) = 1.714, Agriculture, Multidisciplinary, 7/56).

Maja Lj. Bulatović, Rakin Marica B., Mojović Ljiljana V., Nikolić Svetlana B., Vukašinović-Sekulić Maja S., Đukić-Vuković Aleksandra P. (2014). Improvement of production performance of functional fermented whey-based beverage, *Chemical Industry and Chemical Engineering Quarterly*, 20 (1) 1-8 (ISSN 1451-9372) (IF (2014) = 0.892, Chemistry, Applied, 48/72).

Maja Lj. Bulatović, Tanja Ž. Krunić, Maja S. Vukašinović-Sekulić, Danica B. Zarić, Marica B. Rakin (2014). Quality attributes of a fermented whey-based beverage enriched with milk and a probiotic strain, *RSC Advances*, 4 (98), 55503–55510. (ISSN 2046-2069) (IF (2014) = 3.840, Chemistry, Multidisciplinary, 33/157).

Maja Lj. Bulatović, Marica B. Rakin, Maja S. Vukašinović-Sekulić, Ljiljana V. Mojović, Tanja Ž. Krunić (2014). Effect of nutrient supplements on growth and viability of *Lactobacillus johnsonii* NRRL B-2178 in whey, *International Dairy Journal*, 34 (1) 109-115. (ISSN 0958-6946) (IF (2013) = 2.297, Food Science and Technology, 32/122).

In: The Diversified Benefits of Cocoa … ISBN: 978-1-53613-258-8
Editor: Bonifacia Zayas Espinal

Chapter 4

COCOA POWDER DRINK MIX: THE APPLICATION OF A NON-INVASIVE NIR SPECTROSCOPY TECHNIQUE FOR THE FAST IN-LINE DETERMINATION OF MIXTURE COMPOSITION

Maja Benković*[*], *Davor Valinger, Tamara Jurina, Ana Jurinjak Tušek and Jasenka Gajdoš Kljusurić
Faculty of Food Technology and Biotechnology, University of Zagreb, Zagreb, Croatia

ABSTRACT

Tropical tree *Theobroma cacao* L. is a source of one of the most highly appreciated raw materials in the food industry: cocoa beans. Raw cocoa beans are used for the production of chocolate, cocoa mass, cocoa butter and cocoa powder. Cocoa powder is a product which is usually lower priced in comparison to cocoa butter and chocolate, but it is further

[*] Corresponding Author Email: mbenkovic@pbf.hr.

used in many more foodstuffs as a raw material. One of the widely used product which includes cocoa powder as a raw material is cocoa powder drink mix. It is produced by mixing cocoa powder with sugar, usually in a 30:70% ratio. Besides cocoa and sugar, milk powder and instantizers (soy lecithin) are added to the mixtures prior to agglomeration and drying. In order to achieve and maintain a constant mixture composition, it is of great importance to use adequate mixing equipment, as well as to monitor the mixing process at all stages. Since industry mixing devices contain large tonnages of powdered material, constant sampling and mixture composition analysis can be somewhat difficult and time consuming. Because of that, the need for a fast, in-line, build-in mixture composition monitoring technique arises. Near infrared spectroscopy (NIRs) is a valuable analytical tool in agriculture and the food industry for rapid and accurate analysis of certain product constituents (Kays et al., 1997). In addition to being rapid, NIRs requires very little sample preparation, requires no chemicals, and produces no chemical waste (Kays et al., 2002).

In this work, the objectives were to explore the potential of NIR usage for monitoring the cocoa powder and sugar mixtures composition. Cocoa powder and powdered sugar mixtures were produced in different ratios (0 – 100%) with the aim of creating a calibration curve based on the NIR spectra recorded for each mixture. The recorded NIR spectra were then subjected to the Principal Component Analysis (PCA) to obtain the factor values of principal components. The factor values were then plotted in relation to the mixture composition to obtain a calibration curve. The Partial Least Squares regression (PLS) was performed to obtain and validate models based on which the concentration of sugar at any given moment in the mixture can be quantitatively and/or qualitatively predicted. Based on the obtained results, NIRs analysis in combination with the PCA and the PLS has proved to be an adequate, fast and non-invasive method for the cocoa powder drink mix composition prediction.

Keywords: cocoa, sugar, NIR spectroscopy, principal component analysis, partial least squares regression

INTRODUCTION

Nowadays, cocoa is one of the most appreciated food commodity, available in a wide variety of products and used in numerous applications

(Belščak-Cvitanović et al., 2010, de Muijnck, 2005). It owes its popularity to a distinctive and widely acceptable flavour, as well as the proven health benefits of cocoa antioxidants, which mostly include polyphenolic compounds such as proanthocyanidins (58–65%), catechins (29–38%) and anthocyanins (1.7–4.0%) (Biehl and Ziegleder, 2003). Cocoa is often added to different desserts, cakes, chocolate milk, and breakfast cereals in the form of cocoa powder. Basic raw material for cocoa powder production is the cocoa bean, harvested from the cocoa tree (*Theobroma cacao* L.). Modern breeding methods have now led to the development of trees of a standard around 3 m height, to allow easy hand harvesting (de Muijnck, 2005). After harvesting, pods are broken open to remove the beans which are then further subjected to fermentation, which is critical for colour and aroma development. After fermentation, the beans are dried and used for further processing which includes roasting, grinding and alkalizing of the cocoa beans to obtain cocoa nibs and cocoa mass. The cocoa mass contains 50 to 55% of cocoa butter and is subjected to pressing to separate the cocoa butter from the rest of the mass, or the so called press cake. The press cake is then ground, blended and cooled to obtain cocoa powder. The cocoa powder is then packaged in 25 kg paper bags and shipped for industrial uses, or lecithin and stabilizers are added to obtain the lecithinated or stabilized cocoa powders for specific uses which include dairy products, ice-crams, frozen desserts, confectionery products, syrups and instant drink mixes. Instantized cocoa powder drink mix is a very popular cocoa product today, since it requires very little effort to reconstitute and prepare. Recent market analysis of cocoa and chocolate instant products in the USA has shown sales revenues of 893 million dollars and is forecast to increase at an annualized rate of 2.1% over the 10 years to 2022 (IBISWorld, 2017). Adding a regular cocoa powder to milk or water and stirring will give an unattractive looking product, with lumps formed on the surface of the solvent due to the hydrophobic character of the cocoa solids. To avoid lump formation and to improve reconstitution properties, various treatments such as lecithinization, stabilization and agglomeration are conducted. Instant powders with good reconstitution properties are usually produced by agglomeration (Galet at al., 2004; Vu et al., 2003, Benković et al., 2015). Agglomeration or particle size enlargement can be performed using different processes: spray drying, tumble growth, fluid bed drying, freeze drying or extrusion (Barbosa-Canovas et al., 2005). The first stages of the instant cocoa powder production include dry mixing of the basic ingredients: cocoa powder and sugar, usually in the 30:70 ratio. These are later supplemented with milk powder, aromas and stabilizers and subjected to agglomeration using binders to promote formation of porous structures with good dispersibility, wettability and solubility. In order to achieve and maintain constant mixture composition, it is of great importance to use adequate mixing equipment, as well as to monitor the mixing process at all stages. Since industry mixing devices contain large tonnages of powdered material, constant sampling and mixture composition analysis can be somewhat difficult and time consuming. Because of that, the need for a fast, in-line, build-in mixture composition monitoring technique arises. The Near Infrared Spectroscopy (NIRS) covers the transition from the visible spectral range to the mid-infrared region. In the area of NIR (800–2500 nm, respectively 12821–4000 cm^{-1}) mainly vibrations of the –CH, –OH, –SH and –NH bonds are observed (Roggo et al., 2007). The NIRS is generally preferred as an analysis method because of its speed, reliability and non-destructiveness to the tested sample (Roggo et al., 2007, Gajdoš-Kljusurić et al., 2015, Valinger et al., 2011, Porep et al., 2015). Due to its proven reliability and speed in multi-constituent monitoring and control, and the availability of fibre optic probes, the NIR reflectance is receiving renewed interest as a potential on-line sensor for particle size. Because only one sensor is needed to monitor both chemical constituents and particle size, the NIR reflectance has advantages in powder analysis over other methods (Pasikatan et al., 2001).

Previous research on application of the NIR spectroscopy to monitor mixture composition and the mixing process of powders is well documented. e.g., Sekulic et al. (1998) used a NIRSystems 6500 monochromator coupled with the blending vessel by means of a fibre-optic to monitor the mixing process of pharmaceutical powders. They emphasize the importance of developing small scale synthetic blends of targeted composition to construct a profile of homogeneity of pharmaceutical blends. Pasikatan et al. (2001) stated that the cross-sensitivity of the NIR reflectance to the particle size of powders and granular materials has not been fully exploited for particle size estimation and emphasize the importance of proper sampling, reference method selection and chemometric models used for data analysis. Lai et al. (2001) used bulk density of the mixtures as a reference value, but argued that the NIRS usually produces a weak signal and that its sensitivity is limited to the drug mixtures where the drug content is below 1% w/w. Koller and co-workers (2011) applied the NIRS for the continuous quantitative in-line monitoring of pharmaceutical powder mixing of binary mixtures in a bladed mixer and used powder flowability data for model development and the UV/VIS measurements as validation data. They concluded that the NIR spectroscopy is a feasible method for monitoring blending dynamics at a certain location of a blender or mixer. Liew at al. (2010) also concluded that the NIRS can be used to monitor the relative distribution of individual blend components in real time and to assess the performance of a bin blender for mixing of the cohesive multi-component powder blends during development and production. In the food industry, moisture content can also be used to develop mathematical models on reflectance and adsorption spectra with the corresponding moisture content values (Heman and Hsieh, 2016).

Furthermore, application of the NIRS has increased due to development of new methods and chemometric techniques allowing data treatment. Chemometrics includes mathematical and statistical methods for the selection of optimal experimental procedure and data treatment. The chemometric methods used for the NIR spectra analysis are 1) mathematical pre-treatments, 2) classification methods and 3) regression methods (Roggo et al., 2007). In this work, the objective was to explore the potential of NIRs for monitoring the cocoa powder and sugar mixtures composition. The cocoa powder and powdered sugar mixtures were produced in different ratios (0 – 100%) with the aim of creating a calibration curve based on the NIR spectra recorded for each mixture. The recorded NIR spectra were then subjected to the Principal Component Analysis (PCA), as a chemometric classification method, to obtain the factor values of principal components. The factor values were then plotted in relation to the mixture composition to obtain a calibration curve. The Partial Least Squares regression (PLS) was performed to obtain and to validate models based on which the concentration of sugar at any given moment in the mixture can be quantitatively and/or qualitatively predicted.

ABOUT OUR RESEARCH

Materials

The cocoa powder with 10-12% fat and crystalline sucrose were supplied by Kraš (Zagreb, Croatia). The crystalline sucrose was ground for 30 seconds to obtain a powdered form prior to mixture formulation using the IKA Control Tube Mill (IKA Werke, Germany).

Mixture Formulation

The cocoa and sugar powder mixtures were formulated with varying sugar contents ranging from 0 to 100% in regular intervals of 5%, reaching the final number of 21 mixtures. The powders were mixed in a Turbula mixer (Willy A. Bachofen Maschienenfabrik, Muttenz, Switzerland) at 40 rpm for 5 minutes to obtain a homogenous blend. The samples were then transferred to plastic containers and stored at ambient conditions ($T = 19°C$ and $RH = 45\%$) until further analysis.

Particle Size Distribution

The particle size distribution was analysed by a laser diffraction particle sizing method. Laser diffraction measures the particle size distributions by measuring the angular variation in intensity of light scattered as a laser beam passes through a dispersed particulate sample. Large particles scatter light at small angles relative to the laser beam and small particles scatter light at large angles. The angular scattering intensity data is then analysed to calculate the size of the particles responsible for creating the scattering pattern, using the Mie theory of light scattering. The Malvern Mastersizer 2000 instrument coupled with the Scirocco 200 dry dispersion unit (Malvern Instruments, Worchestershire United Kingdom) was used to perform the analyses. The pressure was set at 1.5 bars and the feed rate adjusted in the range from 50 – 70%, based on the observed behavior of the sample in feed unit. The laser obscuration limits were set to 2 – 6%, as listed in the dry powder measurement guidelines (Malvern Instruments, 2007). Experimental data for 8 parameters were further used for data analysis: d(0.1), d(0.5), d(0.9), uniformity, D[3, 2], D[4, 3], span and specific surface area. All measurements were done in triplicate.

Moisture Content

The moisture content was determined according to a standardized AOAC gravimetric method, as the difference in mass of the original sample and the sample dried for 3 h at 105°C (AOAC, 1990).

NIR Spectra

The NIR spectra were recorded using the instrument Control Development, Inc., NIR-128-1.7-USB/6.25/50µm, with installed Control Development software Spec32, in the near infrared range from 904 - 1699 nm. The powder mixtures were placed in a dry material sample holder and the optical probe was reclined to the sample surface. To obtain the highest reflectance, the distance between the entrance face of the cable and the sample surface was minimized (Bičanić et al., 2015). No mechanical or chemical treatment of the samples was needed prior to the NIRS measurements. Ten measurements were taken for each sample and the raw data was further used for the Principal Component Analysis (PCA) and the Partial Linear Regression (PLS) modelling.

Data Analysis and Chemometrics

The technique of NIR spectroscopy is based on the electromagnetic absorption at the near-infrared region but the spectral analysis has to be assisted with various chemometric techniques (Ding and Xu, 1999; Alishahi et al., 2010, Valinger et al., 2011).

The Principal Component Analysis (PCA)

The Principal Component Analysis (PCA) is a widely used multivariate statistical analytical technique. It is applied to data to reduce the set of dependent variables (attributes) to a smaller set of underlying variables (called factors) based on patterns of correlation among the original variables (Puri et al., 2016). The PCA also represents the pattern of similarity of the observations and the variables by displaying them as points in maps (Jackson, 1991, Jolliffe, 2002). Data obtained by the NIRS were used to perform the Principal Component Analysis (PCA by means of the statistical software StatSoft STATISTICA v. 10 (StatSoft Inc., USA).

The Partial Linear Regression (PLS)

The PLS regression bears some similarities to the PCA, with a difference that it finds a linear regression model by projecting predicted and observed values to a new space. It is used to find the relations between two sets of data matrices (X and Y) and/or to predict the Y from the X and to describe their common structure (Abdi, 2003).

The PLS regression was performed for the sugar content, the moisture content and the particle size distribution parameters (separately) in relation to the collected NIR spectra.

General outline of the PLS models used is as shown in Eq. 1:

$$Y = b + a \cdot X_i \tag{1}$$

where Y is the observed parameter (the sugar content, the moisture content and the particle size distribution parameters), b is the regression coefficient for the intercept and X_i is the wavelength of the spectrum (I = 904-1699 nm) and u is a regression coefficient computed from the data.

The model adequacy was assessed by the determination coefficients (R^2) and the root mean squared error (RMSE), calculated according to the equation 2.

$$RMSE = \sqrt{\frac{\sum_{i=1}^{n}(y_i - \hat{y_i})^2}{n}} \quad (2)$$

where y_i is the observed value for i-th observation, $\hat{y}_i$ is the predicted valuc, while n is the number of measurements.

Modelling was performed using The Unscrambler X 10.5. trial version (CAMO Software, Oslo, Norway).

RESULTS AND DISCUSSION

The Experimental Data

The cocoa powder and sugar mixtures were produced with an aim of establishing the suitability of the NIRS technique for a fast, in line monitoring of mixture composition. The mixture composition, the moisture content and the particle size distribution raw data are shown in Table 1.

The first column in Table 1 shows the amount of sugar added to the cocoa/sugar mixtures, meaning that the first row (sugar amount 0%) actually represents the parameters determined for the cocoa powder as raw material and the last row (sugar amount 100%) the parameters for the powdered sugar as a raw material. All the rows in-between represent data for the binary cocoa/sugar mixtures. It is visible from Table 1 that the cocoa powder had the highest moisture content (4.71 ± 0.12%) and the powdered sugar the lowest (0.07 ± 0.00%). An increase of the sugar content in the mixtures led to a decrease in the moisture content, since the raw material with higher moisture content (cocoa powder) was gradually replaced by the one with the lower moisture content (sugar). The particle size parameters exhibited an opposite trend: the higher sugar content of the mixtures also meant higher values of d(0.1), d(0.5), d(0.9), D[3, 2] and D[4, 3]. Sugar was characterized by the median diameter d(0.5) = 121.21 ± 2.79 μm, the volume weighted mean D[4, 3] = 157.85 ± 2.47 μm and the surface weighted mean D[3, 2] = 34.46 ± 1.18 μm. In comparison to the sugar powder, the cocoa powder had significantly lower particle sizes: d(0.5) = 14.64 ± 0.41 μm, D[4, 3] = 24.78 ± 0.26 μm and D[3, 2] = 9.11 ± 0.35 μm. By increasing the amount of sugar in the mixtures, an increase of mixtures' d(0.1), d(0.5), d(0.9), D[3, 2] and D[4, 3] subsequently followed. The span and the uniformity data, although derived from the particle size distribution parameters, exhibited a rise to a maximal value of span = 9.05 ± 0.08 and uniformity = 2.56 ± 0.02 detected for the mixture with 35% sugar. After reaching a maximum, values for the uniformity and the span for the mixtures containing more than 35% sugar started decreasing. Although one would expect the maximum to be reached at 50% sugar and 50% cocoa in a binary mixture, the maximum is shifted towards lower percentages of sugar in the mixtures due to the volume distributions of the particles. The specific surface area results are in accordance to the literature data: higher particle sizes exhibited lower specific surface areas (Barbosa-Canovas et al., 2005).

In order to show the relations and the possibility of the NIRS usage for detection of the mixture composition, the NIR spectra were recorded for all 21 samples. The data is shown in Figure 1.

As seen in Figure 1, most of the spectra parts appear similar throughout all the wavelengths, even though the baseline shift can be noticed. The difference in baseline variation can occur as one of the scattering effect in the NIRS which usually consists of the following: 1) the additive effect, which is reflected as a baseline off set (simple baseline shift) in the NIR spectra; 2) the multiplicative effect (pure) that scales the entire spectrum by a given factor, for example due to the path length differences; 3) the wavelength-dependent baseline variation, where the degree of baseline shift varies with the wavelength. The biggest differences among the raw spectral data can be observed in the area ranging from 1440 to 1460 nm, which is a part of the spectra where the absorption band of carbohydrates is the most pronounced (Toledo et al., 2017).

Scattering is a complex phenomenon and could be linear or non-linear, making it difficult to minimize or remove. Martens et al. (2003) pointed out that scattering correction can be done through implicit and explicit scatter correction. The implicit correction using the multivariate calibration, can to some extent compensate for unknown scatter variation at the expense of the increasing model complexity (more components). However, the bilinear methods (additive first-order approximation), e.g., PCA/PLS are not as effective for accommodating the impact of multiplicative scattering. To minimize the complex baseline variation more effectively, explicit scattering corrections through appropriate data-pre-treatment are needed (Huang et al., 2010).

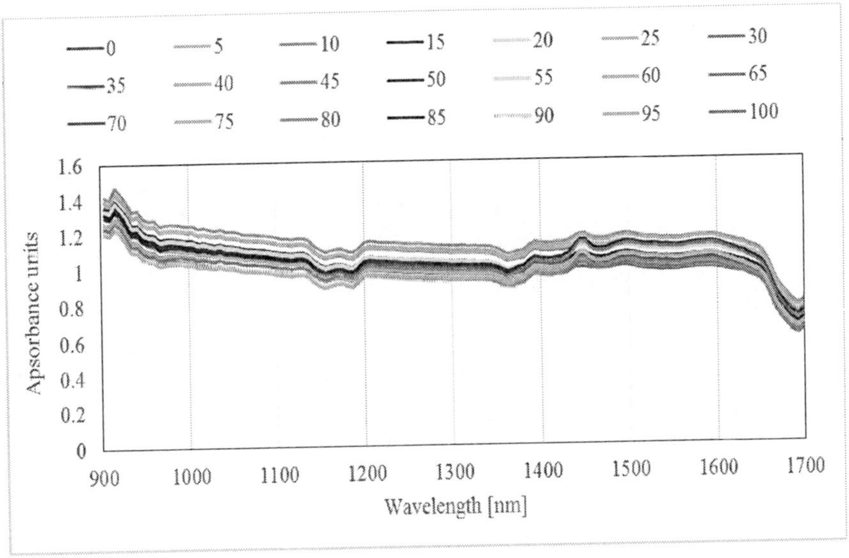

Figure 1. The NIR spectra of the 21 cocoa/sugar mixtures.

Table 1. The experimental data for the mixture composition, the moisture content and the particle size distribution

Sugar amount [%]	Moisture content [%]	d (0.1) [μm]	d (0.5) [μm]	d (0.9) [μm]	D [3, 2] [μm]	D [4, 3] [μm]	Span [/]	Uniformity [/]	Specific surface area [m²/g]
0	4.71 ± 0.12	5.52 ± 0.22	14.64 ± 0.41	57.41 ± 1.12	9.11 ± 0.35	24.78 ± 0.26	3.54 ± 0.16	1.11 ± 0.04	0.66 ± 0.002
5	4.39 ± 0.08	5.40 ± 0.01	14.84 ± 0.13	68.12 ± 1.17	9.05 ± 0.05	28.37 ± 0.44	4.22 ± 0.04	1.34 ± 0.02	0.66 ± 0.001
10	4.20 ± 0.08	5.49 ± 0.02	15.67 ± 0.07	86.68 ± 3.13	9.34 ± 0.02	34.18 ± 0.89	5.18 ± 0.18	1.62 ± 0.05	0.64 ± 0.002
15	3.91 ± 0.05	5.59 ± 0.01	16.81 ± 0.09	120.82 ± 1.30	9.80 ± 0.03	42.49 ± 0.45	6.85 ± 0.05	1.98 ± 0.01	0.61 ± 0.020
20	3.67 ± 0.10	5.80 ± 0.14	18.95 ± 0.89	159.86 ± 9.41	10.51 ± 0.14	52.04 ± 2.58	8.13 ± 0.10	2.23 ± 0.01	0.57 ± 0.002
25	3.61 ± 0.03	5.84 ± 0.29	20.25 ± 0.29	176.24 ± 4.80	10.76 ± 0.06	56.86 ± 1.68	8.41 ± 0.11	2.31 ± 0.05	0.59 ± 0.005
30	3.33 ± .0.03	5.94 ± 0.03	21.85 ± 0.65	197.64 ± 9.32	11.09 ± 0.10	63.10 ± 2.79	8.77 ± 0.18	2.40 ± 0.05	0.54 ± 0.002
35	3.09 ± 0.01	6.49 ± 0.01	26.69 ± 0.36	248.12 ± 4.77	16.27 ± 0.08	80.36 ± 1.49	9.05 ± 0.08	2.56 ± 0.02	0.37 ± 0.002
40	2.78 ± 0.05	6.84 ± 0.13	30.07 ± 0.70	262.66 ± 6.23	16.87 ± 0.89	87.00 ± 1.84	8.51 ± 0.06	2.46 ± 0.00	0.36 ± 0.020
45	2.56 ± 0.04	6.57 ± 0.11	33.57 ± 3.01	282.96 ± 2.53	15.81 ± 0.21	94.42 ± 11.12	8.23 ± 1.47	2.41 ± 0.31	0.38 ± 0.005
50	2.26 ± 0.01	6.79 ± 0.09	44.27 ± 2.12	293.69 ± 6.39	16.29 ± 0.34	104.18 ± 3.01	6.41 ± 0.17	1.97 ± 0.04	0.37 ± 0.007
55	1.91 ± 0.01	7.09 ± 0.11	54.91 ± 4.05	303.46 ± 6.46	17.55 ± 0.51	111.35 ± 3.60	5.40 ± 0.30	1.69 ± 0.08	0.34 ± 0.009
60	1.73 ± 0.03	7.47 ± 0.21	67.29 ± 2.86	320.61 ± 5.11	19.10 ± 0.89	121.37 ± 4.11	4.65 ± 0.43	1.48 ± 0.12	0.31 ± 0.015
65	1.46 ± 0.10	7.77 ± 0.23	74.34 ± 1.08	331.77 ± 6.74	20.19 ± 1.13	127.15 ± 7.41	4.36 ± 0.23	1.38 ± 0.07	0.30 ± 0.016
70	1.34 ± 0.10	7.84 ± 0.39	75.31 ± 1.04	322.10 ± 2.46	20.53 ± 1.73	125.14 ± 4.18	4.17 ± 0.60	1.33 ± 0.18	0.29 ± 0.025
75	0.52 ± 0.02	8.87 ± 0.19	96.97 ± 3.26	354.37 ± 3.11	24.32 ± 1.10	142.95 ± 1.29	3.57 ± 0.14	1.14 ± 0.04	0.25 ± 0.011
80	0.88 ± 0.02	9.63 ± 0.36	107.18 ± 3.21	363.03 ± 11.12	26.89 ± 1.16	150.20 ± 3.19	3.30 ± 0.15	1.05 ± 0.05	0.22 ± 0.010
85	0.36 ± 0.01	10.55 ± 0.22	116.36 ± 2.36	376.77 ± 19.49	29.30 ± 0.35	158.42 ± 5.74	3.15 ± 0.16	1.00 ± 0.05	0.20 ± 0.205
90	0.22 ± 0.01	11.45 ± 0.54	117.14 ± 2.08	362.55 ± 11.78	30.08 ± 1.37	155.39 ± 2.32	3.00 ± 0.15	0.95 ± 0.05	0.20 ± 0.010
95	0.08 ± 0.00	12.40 ± 0.80	115.79 ± 1.70	365.67 ± 12.91	30.35 ± 2.25	155.88 ± 6.75	3.05 ± 0.14	0.96 ± 0.04	0.20 ± 0.014
100	0.07 ± 0.00	15.21 ± 0.37	121.21 ± 2.79	360.32 ± 5.56	34.46 ± 1.18	157.85 ± 2.47	2.85 ± 0.06	0.89 ± 0.02	0.17 ± 0.006

The Principal Component Analysis (PCA)

According to the previous research, the NIRS can be used for characterization of the powder mixing process (Sekulic et al., 1998, Liew et al., 2010) detection of moisture content in foods (Heman and Hsieh, 2016), lycopene content in tomato (Bičanić et al., 2015) and particle size characterization of powders (Pasikatan et al., 2001). In this study, the mixture composition, the moisture content and the particle size distribution were used to develop the calibration models. After obtaining the NIR spectra, the raw data was processed and used for the PCA analysis to see whether the samples can be divided into distinctive groups based on their spectral data. Results of the PCA analysis are shown in Figure 2.

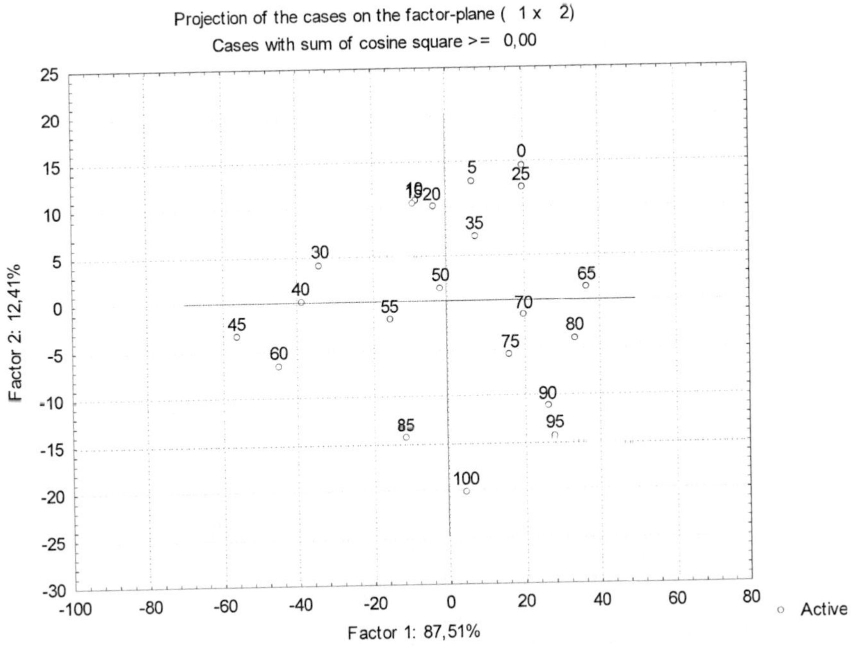

Figure 2. Distribution (2D) of the cases on the factor plane of the cocoa/sugar mixtures with different sugar contents (0-100%) based on the PCA analysis.

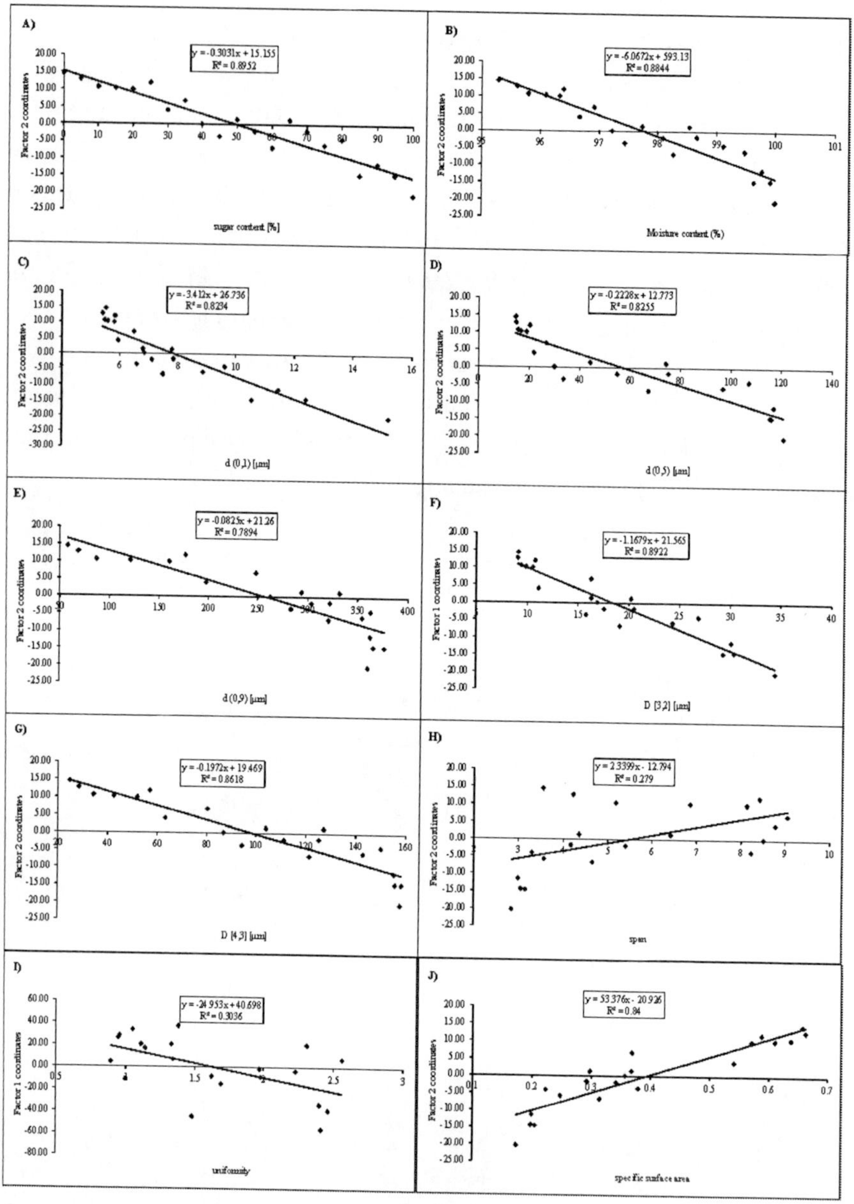

Figure 3. The correlations between the PCs and A) sugar content, B) moisture content, C) d(0.1), D) d(0.5), E) d(0.9), F) D[3, 2], G) D[4, 3], H) span, I) uniformity and J) specific surface area of the cocoa/sugar mixtures.

The PCA analysis transforms operational variables data into new orthogonal variables known as the principal components, which are linear combinations of the original data. The purpose of the PCA is to derive a small number of independent linear combinations of a set of variables measured from a large data set (Toledo et al., 2017). The data presented in Figure 2 show the separation of a large dataset collected by the NIRS. It can be noticed that the raw materials, sugar (sample name "100") and cocoa (sample name "0") are clearly separated and situated in the first (cocoa) and the fourth quadrant (sugar). All of the mixtures are scattered throughout all four quadrants. The samples with lower sugar concentrations are mostly based in the 1st and 2nd quadrant, closer to the cocoa sample "0", while the mixtures containing higher amounts of sugar spread through the 3rd and 4th quadrant, closer to the sugar sample ("100"). Based on the factor values, it can be seen that the factors 1 and 2 together explained 99.92% variances between the samples, indicating excellent separation of the samples by the principal components (PCs). In order to determine which factor from the PCA analysis correlated to the sugar content, the moisture content and the particle size, the principal components (PCs) were plotted against the experimentally determined values of the mentioned parameters. The plots are shown in Figure 3.

The factors obtained by the PCA analysis were further used to determine the correlations between all parameters obtained from the experiment. The correlation between the sugar content and the Factor 2 (PC2) obtained by the PCA analysis is shown in Figure 3A. The value of the determination coefficient (R^2 = 0.8952) suggested the existence of a strong bond between the PC 2 and the sugar content, indicating that the PCA analysis was able to distinguish the mixtures based on their composition. However, the PC 2 coordinates were also correlated to the moisture content (R^2 = 0.8844, Figure 3B), d (0.1) (R^2 = 0.8234, Figure 3C), d (0.5) (R^2 = 0.8255, Figure 3D), d (0.9) (R^2 = 0.7894, Figure 3E), D [4, 3] (R^2 = 0.8618, Figure 3G), and the specific surface area (R^2 = 0.8400, Figure 2J), indicating that all of the listed parameters can be used as reference values for creation of the NIRS calibration model. The only exception was seen for D [3, 2], which exhibited the highest correlation to the PC 1 (R^2 = 0.8922, Figure 3F). As seen in Figures 3H and 3I, the span and the uniformity did not show strong correlations to the PCA factors (R^2 = 0.2790 for span and R^2 = 0.3036 for uniformity, respectively). The span and the uniformity were discussed before :their values increased to the maximum reached for the mixture containing 35% of sugar and then decreased again for mixtures containing 40 to 100% sugar. This trend cannot be described by any linear correlation.

The Partial Least Squares Regression (PLS)

The PLS regression is another multivariate method whose use as a chemometric tool has been previously documented (Pasikatan et al., 2001, Mabood et al., 2017). In comparison to the PCA analysis, although there are some similarities, the PLS is aimed to find some relation between the response and the independent variables, while the PCA is mainly used to detect the similarities or the differences between the used variables. In this study, the PLS regression was performed to correlate the sugar content, the moisture content and the particle size parameters to the NIR spectra. All variables were correlated to the NIR spectral data separately, resulting in 10 generated models. Results are shown in Table 2.

As seen in Table 2, the PLS models were developed for all experimental parameters, but in comparison to the PCA, part of the data was used for calibration and another part for model validation. Generally speaking, when using the PLS regression, the R^2 values higher than 0.90 indicate an excellent fit of the proposed model to the experimental data. All models had the determination coefficient values higher than 0.9, with an exception of the span and theuniformity for the model validation. The determination coefficients of the calibration dataset were slightly higher than those for the validation dataset in all cases. As mentioned before, the span and the uniformity validation models were the only exception with lower R^2 values, which was an indication that, although the calibration model was generated by the PLS regression, it cannot be used in further predictions due to low values of the R^2 (R^2 = 0.6486 for span and R^2 = 0.6390 for uniformity) for validation.

As for all the other parameters, the highest R^2 was determined for the sugar content (R^2 = 0.9877 and 0.9817), followed by the moisture content (R^2 = 0.9858 and 0.9784) and the D [3, 2] (R^2 = 0.9814 and 0.9791). These results indicate that the PLS can successfully be applied for the calibration and validation of large datasets derived from the NIR spectra and that the sugar content, the moisture content and laser diffraction particle sizing can be used as the reference methods for the NIR calibration.

Table 2. The PLS model coefficients for 21 cocoa/powder mixtures containing different amounts of sugar

Variable	Data set	Slope	Offset	RMSE	R^2
Sugar content [%]	Calibration	0.9877	0.6147	3.3569	0.9877
	Validation	0.9807	1.1195	4.2611	0.9817
Moisture [%]	Calibration	0.9858	0.0318	0.1793	0.9858
	Validation	0.9836	0.0199	0.2354	0.9784
d (0.1) [μm]	Calibration	0.9811	0.1480	0.3545	0.9811
	Validation	0.9259	0.5869	0.5858	0.9571
d (0.5) [μm]	Calibration	0.9652	1.9921	7.3736	0.9652
	Validation	0.9666	2.0380	8.9444	0.9528
d (0.9) [μm]	Calibration	0.9329	17.2923	27.0652	0.9329
	Validation	0.9190	21.6916	32.6582	0.9115
D [3, 2] [μm]	Calibration	0.9814	0.3423	1.0681	0.9814
	Validation	0.9795	0.3812	1.2323	0.9791
D [4, 3] [μm]	Calibration	0.9718	2.7800	7.6623	0.9718
	Validation	0.9715	3.1924	9.8264	0.9577
Span [/]	Calibration	0.9730	0.1477	0.3599	0.9730
	Validation	0.8473	0.6564	1.4067	0.6486
Specific surface area [m²/g]	Calibration	0.9263	0.0289	0.0452	0.9263
	Validation	0.9221	0.0293	0.0536	0.9081
Uniformity [/]	Calibration	0.9793	0.0338	0.0818	0.9793
	Validation	0.8777	0.1629	0.3438	0.6390

CONCLUSION

In this chapter, the objectives were to explore the potential of the NIR usage for monitoring the cocoa powder and sugar mixtures composition. The cocoa powder and powdered sugar mixtures were produced in different ratios (0 – 100%) with the aim of creating a calibration curve based on the NIR spectra recorded for each mixture. The PCA analysis proved to be useful for distinguishing the differences and similarities between the NIR spectra of all the samples. The factors calculated using the PCA analysis (PCs) can successfully be used to create the calibration models based on the sugar content, the moisture content, d(0.1), d(0.5), d(0.9), D[3, 2], D[4, 3] and the specific surface area of the mixtures. The PLS regression can also be used successfully for calibration of the same parameters listed above with high determination coefficients ($R^2 > 0.90$, respectively). Based on the obtained results, the NIRs analysis in combination with the PCA and the PLS has proved to be an adequate, fast and non-invasive method for the cocoa powder drink mix composition prediction.

REFERENCES

Abdi, H. (2003). PLS-Regression; Multivariate analysis. In M. Lewis-Beck, A. Bryman & T. Futing (Eds.), *Encyclopedia for research methods for the social sciences*, (1st edition, pp. 792-795). Thousand Oaks, USA, SAGE Publications.

Alishahi, A., Farahmanda, H., Prietob, N. & Cozzolino, D. (2010). Identification of transgenic foods using NIR spectroscopy: a review. *Spectrochim. Acta A.*, *75*, 1–7.

AOAC International. (1990). *Official Methods of Analysis* (15th edition). Arlington, USA, Association of Official Analytical Chemists.

Barbosa-Canovas, G. V., Ortega-Rivas, E., Pablo, J. & Hong, Y. (2005). *Food Powders: Physical Properties, Processing, and Functionality* (1st edition). New York, USA, Kluwer Academic/Plenum Publishers.

Belščak-Cvitanović, A., Benković, M., Komes, D., Bauman, I., Horžić, D., Dujmić, F. & Matijašec, M. (2010). Physical properties and bioactive constituents of powdered mixtures and drinks prepared with cocoa and various sweeteners. *J. Agric. Food Chem.*, *58*(12), 7187-7195.

Benković, M., Jurinjak Tušek, A., Belščak-Cvitanović, A., Lenart, A., Domian, E., Komes, D. & Bauman, I. (2015). Artificial neural network modelling of changes in physical and chemical properties of cocoa powder mixtures during agglomeration. *LWT - Food Sci. Technol.*, *64*, 140-148.

Bičanić, D., Streza, M., Dóka, O., Valinger, D., Luterotti, S., Ajtony, Z., Kurtanjek, Ž. & Dadarlat, D. (2015). Non-destructive Measurement of Total Carotenoid Content in Processed Tomato Products: Infrared Lock-In Thermography, Near-Infrared Spectroscopy/Chemometrics, and Condensed Phase Laser-Based Photoacoustics—Pilot Study. *Int. J. Thermophys.*, *36*(9), 2380-2388

Biehl, B. & Ziegleder, G. (2003). Cocoa. In: B. Caballero (Ed.), *Encyclopedia of food science and nutrition*, (2nd edition, pp. 1436-1463). New York, USA, Academic Press.

de Muijnck, L. (2005). Cocoa. In C. Onwulata (Ed.), *Encapsuled and Powdered Foods*, (1st edition, pp. 451-473), Boca Raton, USA, CRC Press.

Ding, H. B. & Xu, R. J. (1999). Differentiation of Beef and Kangaroo Meat by Visible/Near Infrared Reflectance Spectroscopy. *Journal Food Sci.*, *64*(5), 814-817.

Gajdoš Kljusurić, J., Benković, M. & Bauman, I. (2015). Classification and processing optimization of barley milk production using NIR spectroscopy, particle size and total dissolved solids analysis. *J. Chem.*, http://dx.doi.org/10.1155/2015/896051.

Galet, L., Vu, T. O., Oulahna, D. & Fages, J. (2004). The wetting behaviour and dispersion rate of cocoa powder in water. *Food Bioprod. Process.*, *82*(C4), 298-303.

Heman, A. & Hsieh, C. (2016). Measurement of moisture content for rough rice by visible and near-infrared (NIR) spectroscopy. *Engineering in Agriculture, Environment and Food*, *9*(3), 280-290.

Huang, J., Romero-Torres, S. & Moshgbar, M. (2010). Practical Considerations in Data Pre-treatment for NIR and Raman Spectroscopy. *Am. Pharm. Rev.*, *13*(6), 116-127.

IBISWorld [online] *Cocoa & Drinking Chocolate Production: Market Research Report*. [2017, October 11th] Available from: https:// www. ibisworld.com/industry-trends/specialized-market-research-reports/ consumer-goods-services/beverage-production/cocoa-drinking-chocolate-production.html.

Jackson, J. E. (1991). *A User's Guide to Principal Components*. New York, USA, John Wiley & Sons.

Jolliffe, I. T. (2002). *Principal Component Analysis* (2nd edition). New York, USA, Springer-Verlag.

Kays, S. E. & Barton, F. E. (2002). Near-infrared analysis of soluble and insoluble dietary fibre fractions of cereal food products. *J. Agric. Food Chem.*, *50*(10), 3024-3029.

Kays, S. E., Barton, F. E., Windham, W. R. & Himmelsbach, D. S. (1997). Prediction of total dietary fiber by Near-infrared reflectance spectroscopy in cereal products containing high sugar and crystalline sugar. *J. Agric. Food Chem.*, *45*(10), 3944–3951.

Koller, D. M., Posch, A., Hörl, G., Voura, C., Radl, S., Urbanetz, N., Fraser, S. D., Tritthart, W., Reiter, F., Schlingmann, M. & Khinast, J. G. (2011). Continuous quantitative monitoring of powder mixing dynamics by near-infrared spectroscopy. *Powder Technol.*, *205*(1–3), 87-96.

Lai, C. K., Holt, D., Leung, J. C. & Cooney, C. L. (2001). Real Time and Noninvasive Monitoring of Dry Powder Blend Homogeneity. *AIChE Journal*, *47*(11), 2618-2622.

Liew, C. V., Karande, A. D. & Heng, P. W. S. (2010). In-line quantification of drug and excipients in cohesive powder blends by near infrared spectroscopy. *Int. J. Pharm.*, *386*, 138-148.

Mabood, F., Jabeen, F., Ahmed, M., Hussain, J., Al Mashaykhi, S. A. A., Al Rubaiey, Z. M. A., Farooq, S., Boqué, R., Ali, L., Hussain, Z., Al-Harrasi, A., Khan, A. L., Naureen, Z., Idrees, M. & Manzoor, S. (2017). Development of new NIR-spectroscopy method combined with multivariate analysis for detection of adulteration in camel milk with goat milk. *Food Chem.*, *221*, 746-750.

Martens, H., Nielsen, J. P. & Engelsen, S. B. (2003). Light scattering and light absorbance separated by extended multiplicative signal correction. Application to near-infrared transmission analysis of powder mixtures. *Anal. Chem.*, *75*(3), 394-404.

Malvern Instruments. (2007). *Sirocco 2000 User manual* (Issue 01). Worchestershire, United Kingdom, Malvern Instruments Ltd.

Pasikatan, M. C., Steele, J. L., Spillman, C. K. & Haque, E. (2001). Near infrared reflectance spectroscopy for online particle size analysis of powders and ground materials. *J. Near Infrared Spec.*, *9*, 153-164.

Porep, J., Kammerer, D. R. & Carle, R. (2015). On-line application of near infrared (NIR) spectroscopy in food production. *Trends Food Sci. Technol.*, *46*(2), *Part A*, 211-230.

Puri, R., Khamrui, K., Khetra, Y., Malhotra, R. & Devraja, H. C. (2016). Quantitative descriptive analysis and principal component analysis for sensory characterization of Indian milk product cham-cham. *J. Food Sci. Technol.*, *53*(2), 1238-1246.

Roggo, Y., Chalus, P., Maurer, L., Lema-Martinez, C., Edmond, A. & Jent, N. (2007). A review of near infrared spectroscopy and chemometrics in pharmaceutical technologies. *J. Pharm. Biomed. Anal.*, *44*(3), 683-700.

Sekulic, S. S., Wakeman, J., Doherty, P. & Hailey, P. A. (1998). Automated system for the on-line monitoring of powder blending processes using near-infrared spectroscopy Part II. Qualitative approaches to blend evaluation. *J. Pharm. Biomed. Anal.*, *17*, 1285-1309.

Toledo, M., Gutiérrez, M. C., Siles, J. A., García-Olmo, J. & Martín, M. A. (2017). Chemometric analysis and NIR spectroscopy to evaluate odorous impact during the composting of different raw materials. *J. Clean. Prod.*, *167*, 154-162.

Valinger, D., Benković, M., Gajdoš Kljusurić, J., Bauman, I. & Kurtanjek, Ž. (2011). Application of NIR spectroscopy for monitoring different particle sizes of sucrose. *Journal on processing and energy in agriculture*, *15*(3), 188-190.

Vu, T. O., Galet, L., Fages, J. & Oulahna, D. (2003). Improving the dispersion kinetics of a cocoa powder by size enlargement. *Powder Technol.*, *130*, 400-406.

In: The Diversified Benefits of Cocoa … ISBN: 978-1-53613-258-8
Editor: Bonifacia Zayas Espinal

Chapter 5

A Technique to Increase Cocoa Pod Production by Enhancing Insect Pollination: Implications for Different Environments

Allen M. Young[1,*] ***and J. Robert Hunter***[2,†]
[1]Department of Invertebrate Zoology,
Milwaukee Public Museum, Milwaukee, WI, US
[2]Herbarium, The University of Wisconsin-Madison,
Madison, WI, US

Abstract

The world's production of cocoa beans is directly dependent on the pollination of cocoa flowers by insects, chiefly midges of the diptern family Ceratopogonidae and to a lesser extent, Cecidomyiidae. Previous

---

* Corresponding Author Email: young@mpm.edu.
† J. Robert Hunter (deceased 1997), the study was done on his property and he assisted with the design of the study and its implementation.

studies in Central America suggested that increases in cocoa pod production is directly related to increased abundance of these pollinators.

In these studies, banana pseudostem pieces placed in piles at the base of cocoa trees, and upon rotting under humid conditions, provided ideal breeding sites for the life cycle of cocoa-pollinating midges. The study presented in this paper is an attempt to confirm the relationship between increased pollinator abundance and higher yields of cocoa pods at one locality in Costa Rica. While cocoa pollination can depend upon a diverse array of ceratopogonid midges, in this study, one species *Forcipomyia youngii*, dominated the samples. The technique used was a four-year "cross over" study in which two plots, each with 25 cocoa trees under a shade canopy of chiefly *Cordia alliodora* "laurel" trees received a treatment of pseudostems twice.

If pod production significantly increased with the treatment, then the treatment was effective when applied to Lot A over two years and when applied to Lot B for another two years. For each plot when given the treatment, an increased abundance of pollinators would result in increased pollination and more pods. Analysis of the data revealed no significant increase in pod production with treatment but trends in that direction.

Yet we conclude that this technique could demonstrate increases in cocoa pod production under different environmental conditions. While the technique is less effective in highly humid conditions as the present study site, it could be far more effective as a means of expanding cocoa growing in drier regions and perhaps in association with water-drip irrigation. Under dry conditions, moist and humid piles of banana pseudostems would provide a suitable breeding niche for pollinators. Under these conditions, a crossover in breeding substrate treatment for pollinating midges is predicted to be far more effective than under uniformly humid conditions.

Thus, there is a potential agronomic benefit is in the application of this technique. It can diversify where cocoa trees are grown, i.e., in drier regions. Therefore, this study warrants replication along an environmental gradient, from very humid to very dry conditions. The use of banana pseudostems to breed pollinating midges is a relatively inexpensive technique using a resource (banana plants) typically available where cocoa is grown. These findings are further discussed in terms of the environmental benefits of growing cocoa as a cash crop integrated into agroforestry systems.

INTRODUCTION

The world's production of cocoa beans and products derived from them, such as chocolate and cocoa butter, is dependent upon insect-mediated pollination of flowers in cocoa (*Theobroma cacao*) trees (e.g., Young 1994 and included references). An equally world-wide challenge in producing cocoa beans is that the level of pollination and pollinator abundance tends to be very low in cultivated orchards of cocoa trees (Figure 1). The principal cocoa-pollinating insects are chiefly small-sized midges in the dipteran family Ceratopogonidae and to a lesser extent Cecidomyiidae (Young 1994). Ceratopogonid midges are known to be the principal pollinators in all cocoa-growing regions of the world (e.g., Wellensiek 1932, Cope 1939, Dessart 1961, Winder and Silva 1975). The role of these midges in pollination of cocoa flowers has been well studied in the Neotropics (e.g., Soria 1970, Soria and Wirth 1979, Young 1982, 1983, 1986).

It was previously demonstrated that the consistent application of sliced-up pieces of banana (*Musa*) pseudostems placed in piles beneath cocoa trees in Costa Rica greatly increased the abundance of ceratopogonid immature life stages as contrasted with the more typical breeding substrate of moist leaf litter (Young 1983). That study demonstrated an increase in cocoa pods although not a statistically significant one. Based on these findings, a four-year experiment (1989-1993) was conducted in Costa Rica and designed to further test the pseudostem technique as a means of increasing pollinator abundance which in turn could possibly increase the yield of cocoa beans, something of considerable economic importance to cocoa growers. In this paper, we suggest that the key role of natural pollination of cocoa trees can be potentially enhanced by use of a novel and relatively inexpensive technique that employs a resource readily available in cocoa-growing regions. Such results, under drier conditions in particular, can be a prerequisite for a set of diversified environmental benefits utilizing cocoa trees.

Figure 1. A cocoa grove, Finca Experimental La Lola, Siquirres, Limon Province, Costa Rica.

Materials and Methods

Project Locality

This study was conducted at "Finca La Tigra," situated adjacent to "Finca La Tirimbina" (now the Tirimbina Rainforest Center or Estacion Biologica La Tirimbina) and in the vicinity of the town of La Virgen de Socorro (10° 23' N, 84° 07' W) in the Heredia Province of northeastern Costa Rica. The region is called "the Sarapiqui" (see Young 2017) and consists of a mosaic of farms and premontane tropical rainforest.

Monthly and annual rainfall date from Tirimbina indicates the absence of a pronounced dry season at this locality (Table 1). While generally the dry season occurs during February, March and April at many Costa Rican localities, these data show considerable precipitation during these months at Tirimbina and La Tigra.

Table 1. Monthly total precipitation for Finca La Tirimbina, La Virgen de Socorro, Sarapiqui, Costa Rica for years 1977 to 1993

Year	Jan	Feb	Mar	Apr	May	Jun	Jul	Aug	Sep	Oct	Nov	Dec	Totals
1977	134	145	152	162	214	728	335	353	432	641	426	119	3921
1978	199.5	288	212.6	63	371.5	381.5	594	563	291.5	430	491	410	4295.6
1979	164	178	108	514	235.5	401	449.5	728	307	409	542.5	456.6	4493
1980	255.5	189	79	173.5	291.7	613.5	664.5	364.5	464	224.5	447	764.5	4531.2
1981	438.5	345	251.5	242.5	479	541	376	533	249.5	586.5	952.5	527.5	5522.5
1982	112	77	47.5	66	193	526	1055	602.5	367.5	733.5	457.7	472	4709.7
1983	345	235	275.5	31.5	474.5	267.5	791.5	354.5	293	518	195.5	177	3958.5
1984	278.5	176	66.5	51	311	433	378.5	483.5	359.5	491	439.5	327	3797.5
1985	104.5	230	97.5	118.5	141.5	515.5	34.7	407.5	331.5	270	218.5	185	3043.5
1986	226.5	86.5	290.5	138.5	205	492	457	478	517	255	215	231	3592
1987	273	118	19	257	212	426	518	561	255	457	344	291	3671
1988	n.d.	252	202	71	553	239	317	478	494	488	269	602	
1989	237	104	151	225	342	496	752	309	345	370	432	302	4065
1990	384.7	63.8	328.7	129.8	614.2	575.6	377.8	825.5	363.2	245	406.9	311.8	4627
1991	169.8	278.9	88.2	240.0	467.8	465.1	776.3	624.4	462.9	299.6	375	192	4440.9
1992	105.5	122.9	157.9	168.2	229.6	345.2	607.7	256.2	251.8	416.9	107.4	239.4	3008.7
1993	127	54.9	163.4	75.1	225.5	328.2	474.9	505.5	441.6	374.9	229.8	309.4	3310.2

All numbers are in mms.*

*Data from Station No. 39586 of the Costa Rican Servicio Nacional Metereologico where rainfall was recorded daily (Finca La Tirimbina).

Project Design

During February-March of 1989 a total of twenty-five healthy, mature UF-29 cocoa trees in each of two designated study plots (Lots A and B) situated within a large planting of these trees were selected and tagged. The UF-29 cultivar was chosen for this study since it is self-compatible, thereby enhancing our ability to possibly detect any impact on changes in the availability of pollinators (see below) on the observed levels of pod-set. The site is essentially a L-shaped grove of several hundred UF-29 cocoa trees planted in 1974 from budded material. The site has a shade cover of mainly "laurel" (*Cordia alliodora*) trees. The two study plots are partially separated by a small hill of secondary-growth rainforest (at the juncture of the "L") and with about one-eighth of a mile between them. Care was taken to choose trees away from border areas. Individual trees are tagged with bright yellow plastic number-coded tags and also plastic blue streamers to facilitate locating them for census. The total tagged cocoa trees for this experiment consisted of trees 1-25 in Lot A and 26-50 in Lot B. Soon after the study had begun, it was apparent that one of the chosen trees appeared not to be UF-29 based upon the pigmentation of the flowers and was therefore eliminated. With technical assistance chiefly from Lucrecia Guillen, a forestry research assistant at CATIE (Turrialba) stationed at the adjoining La Tirimbina farm in conjunction with a long-term project on sustainable-yield forestry, the numbers of flowers and pods were recorded once monthly, beginning 31 May 1989. Each of these censuses was exhaustive in that all floral buds, open flowers, cherelles (newly-set pods), intermediate-sized pods, harvestable pods (mature pods), cherelle wilt and diseased pods are recorded for the 49 tagged trees each month.

Project Objectives

1. To measure and assess, on a monthly basis, the levels of flowering, pod-set and production of harvestable (mature) cocoa pods in one "control" and one "experimental" study plot (designated as Lots A and B, respectively).
2. To measure and assess annually the comparative abundance and diversity of cocoa-pollinating insects, chiefly Diptera: Ceratopogonidae and Cecidomyiidae, through the sampling of midge larvae and pupae and rearing of adult midges from this material, between the "control" (Lot A) and "experimental" (Lot B) cocoa study plots, in which the latter study plot received, as a designated treatment, twenty-five sliced sections of *Musa* (banana) pseudostems under each tree every six weeks, to encourage midge breeding.
3. To interpret from the results obtained in (1) and (2) above to determine whether or not the deliberate elevating of pollinator

populations in a cocoa plantation increases both the levels of pod-set and harvestable pods, in relation to flowering and pod losses (cherelle wilt, diseases) in an economically-valuable sustained manner for the cocoa grower.

4. To interpret the findings in terms of cocoa as an environmental friendly cash crop that can be integrated into multi-purpose agroforestry systems and old secondary rainforest. And ultimately these benefits and the role of cocoa in eco-tourism depend upon effective levels of midge-mediated pollination.

Experimental Treatment for Pollinators

Based upon the discovery in previous research in Costa Rica (e.g., Young 1983) that rotted sections of *Musa* pseudostems greatly elevate the abundance of immature stage (larvae and pupae), and presumably adults, of cocoa-pollinating midges by acting as a preferred breeding micro-habitat, the twenty-five tagged cocoa trees in the "experimental" plot (B) in the present study, each received 25 pieces of freshly-cut pseudostem slices once every six weeks. Fresh pseudostem pieces were gathered from a nearby banana farm. They were obtained by chopping with a machete the pseudostems of fallen banana plants. The total 625 pieces are distributed, with cut-face down, beneath each of the tagged cocoa trees, and allowed to decay. The "control" plot (A) does not receive the pseudostem treatment. Naturally-occurring groundcover (leaf litter) is kept intact in both plots. Once in place, the *Musa* pseudostem slices are left undisturbed, with the exception of an annual census to assess the abundance of cocoa-pollinating insects. Pseudostem pieces from previous applications are left to decay.

Lot B remained the experimental plot until October 1991 when the switch or cross-over was performed, making Lot A the experimental plot for the remaining two years of the study. As with Lot B in the first two years of the study, varying numbers of rotted pseudostem pieces were checked for midge immatures in each census. The rationale for applying the pseudostems to the cocoa grove was to create a moist breeding substrate for pollinating midges. If the numbers of midges increased as a result of this treatment, then more flowers would be pollinated leading to higher production of cocoa pods. This experiment continued for four years with both plots receiving the treatment twice.

Census of Pollinating Midges

A first census of cocoa-pollinating insects in the two plots was taken during September of 1989, a period corresponding to the late rainy season in this region of Costa Rica. In the experimental plot (B), varying numbers of pieces in intensely rotted pseudostem slices were collected from beneath each of the tagged trees over several days. The pseudostem pieces at each census were not exhaustively collected. Only a sample of pseudostems were collected and examined for midge immatures. Exhaustive collecting would have likely depleted the potential pollinator population and distort results. The pieces collected corresponded to the July 1, 1989 application, since those deposited in the cocoa on August 15 were still too fresh for midge breeding, as judged by previous observations (Young 1983). In the control plot, several large bags of rotted leaf litter, another known breeding substrate for cocoa-pollinating midges, consisting of about 2500 leaves in all, were also collected and examined for the presence of midge larvae and pupae. These collections of "organic trash" were taken to the field house at La Tirimbina where the contents of each piece of pseudostem and all leaves were checked thoroughly for midge immature stages.

To examine a rotted slice of pseudostem, the piece was gently teased or otherwise broken apart, small pieces at a time, and the presence of midge larvae and pupae detected with a hand lens. Prior experience with censusing midge immatures greatly facilitated recognizing the pertinent insects in the debris. Most of the collected midge larvae and pupae were preserved in small vials of 70% ethanol to serve as research vouchers for subsequent identification to genera and species, with the expert cooperation of Drs. Raymond J. Gagne (Cecidomyiidae) and Willis W. Wirth (Ceratopogonidae), entomologists presently associated with the US

Department of Agriculture (Systematic Entomology Laboratory, Washington, DC) and internationally-recognized authorities on these groups of Diptera.

A small portion of the collected larvae and pupae were kept alive in small vials to rear adult voucher specimens as well. The numbers of larvae and pupae of ceratopogonid and cecidomyiid midges found in the samples of rotted pseudostems and leaf litter were recorded in order to evaluate the degree to which breeding populations were at different levels of abundance between the two study plots. The assumption was made that any observed difference in the abundance of immature stages of the midges between the two plots would also be true for the adult stage. A rough tally sheet in which each kind of midge larva or pupa immediately recognizable as distinctive from others retrieved in the samples was given a tentative "morpho-species" name, enabled us to estimate the approximate number of midge species.

RESULTS

Cocoa Pod Production in Relation to Flowering and Treatment

The average numbers of cocoa flowers (Figure 2) and pods (Figure 3) for the entire study are shown. Also indicated is where the cross-over in experimental treatment took place (Figures 4 and 5). An analysis of variance performed on the flowering and pod data for the two lots did not show any statistically significant differences as to the treatment effect of presence/absence of the banana pseudostems for the 3.5 years of the study (Table 2). There was no significant effect of the banana pseudostem treatment ($p = 0.347$) assuming $p = < .05$. However, when the average number of small pods (larger pods are excluded because of overlaps in counting them from month to month since they take about six months to mature) are examined (Figure 5), there appears to be a trend towards higher pod production with treatment. In interpreting the data of pod-set patterns, it is important to keep in mind that the *Musa* pseudostem treatment began on April 1, 1989. Since about 15-20 days are required for ceratopogonid and cecidomyiid midges to complete their life cycles (egg-to-adult), and assuming a month is the minimal amount of time needed for pseudostem sections to decay sufficiently to attract midges for breeding, the true initial starting time for this study would have occurred at the time of the first "cohort" of adult midges hatched from the distributed pseudostems, that is, on or about May 15, 1989. This would be an appropriate date to expect to detect an increase in pod-set on the May 31, 1989 (first) census of the trees in the experimental plot (B), which in fact we did find. Therefore, it is important to consider "lag effects" between midge hatches and pod-setting patterns in this experiment.

For certain periods when Lot B had the treatment there was a period of increased numbers of small (newly-set) pods, as March-May 1991 (Figure 5). But this pattern was not consistent for the duration of the treatment effect period. A similar trend was noted when Lot A had the treatment (Figure 5). For certain trees in both lots, average total pod production seemed to respond to the treatment (Figure 6). For example, in 1991 a block of trees in Lot A farthest removed from Lot B (trees nos. 1-12) contributed least to the average pod production, and following the cross-over, this pattern is well on its way to reversing itself in 1993 for a block of trees in Lot B, and those farthest removed distance-wise from Lot A trees. Again, it appears as if there is some effect of treatment, but it is not consistent in response to the cross-over experimental design for all trees. Based on these results, 50% of the tagged cocoa trees do not respond differently when treated with banana pseudostems than the other 50% of the trees.

When the data are analyzed using Least Square Means (LSM), pod production almost doubled in Lot B (Table 3), but this cannot be ascribed to an effect of the cross-over experiment. Although the pod data for Lot A exhibits the opposite trend, it is much less pronounced than what was observed for Lot B (Table 3). For some unknown reason, the trees in Lot B are consistently better in cocoa pod production than those in Lot A. Small (new) pods generally represented about 10% of the flower production in this study.

Figure 2. Cocoa flowers.

Figure 3. Small cocoa (cacao) pod.

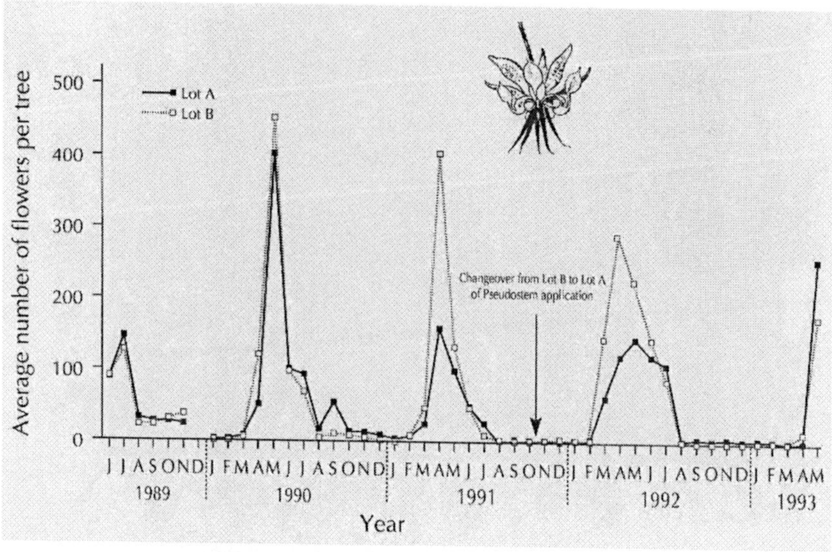

Figure 4. The average number of cocoa flowers in Lots A and B for the duration of this study (Finca La Tigra).

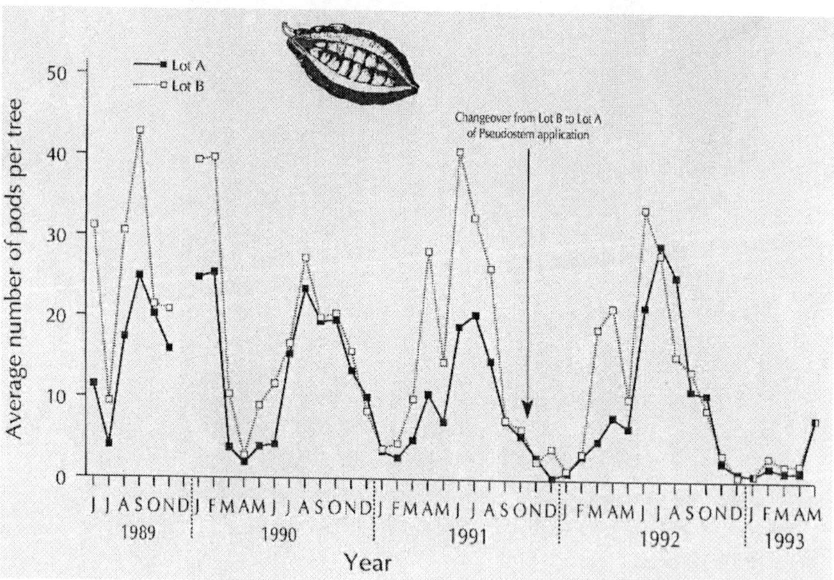

Figure 5. The average number of all pods per tree, monthly, for the duration of this study (Lots A and B).

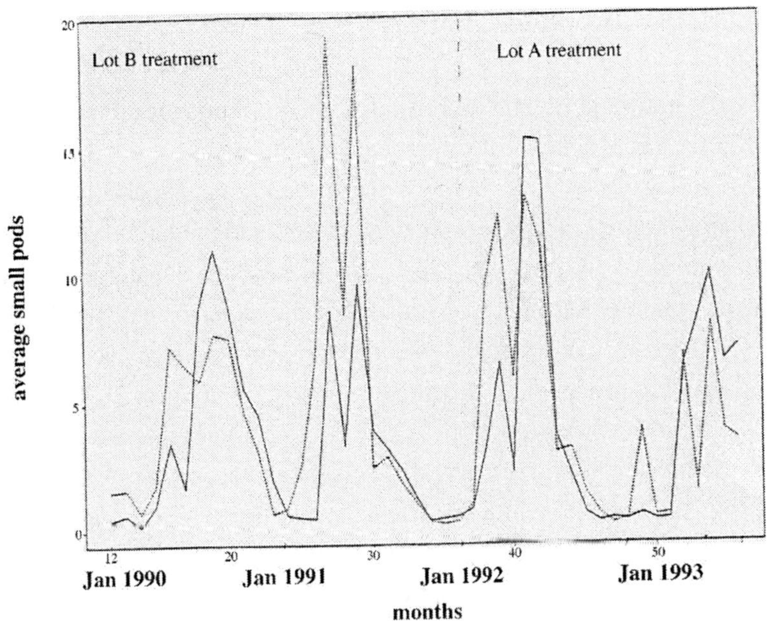

Figure 6. The average number of small (newly-set) cocoa pods on trees in Lots A and B for the duration of the study.

Table 2. Analysis of variance for cocoa pod production for Lots A and B in relation to banana pseudostem treatment

Source	DF	Type III SS	F. Value	Pr > F
Lot	1	1.343	0.96	0.382
Treatment	1	1.580	1.13	0.3472
Low * Treatment	1	0.032	0.02	0.887

Table 3. Least Squares Means (LSM) analysis of flower and pod production between the two lots of UF-29 cocoa trees in relation to banana pseudostem treatment for pollinating midges

Lot	Treatment	Flower LSM	Pod LSM
1	no pseudostems	54.24	4.17
1	with pseudostems	41.35	3.95
2	no pseudostems	58.56	3.75
2	with pseudostems	67.26	6.19

Pollinating Midge Populations

While hardly any immature stages of ceratopogonid midges were found in the large sample of moist leaf litter from Lot A or B when each was the control lot (see below), large numbers, especially those of *Forcipomyia youngii* (a species new to science and named after this author by Dr. Willis W. Wirth), were found in the rotting banana pseudostems (Figure 7) when either Lot A or Lot B was the experimental lot (Figure 8, Tables 4 and 5). Additionally, the pseudostem samples also contained several hundred larvae of unidentified cecidomyiid midges.

A range of 0-7 ceratopogonids and 0-12 cecidomyiids were found for rotted pseudostems, and an overall "percentage occupancy" of 24% was found for midges breeding in pseudostems for the sample collected. Combining the figures for ceratopogonids and cecidomyiids, almost one midge per pseudostem slice in the collected sample occurs on the average for this type of breeding microhabitat. This is a very impressive figure, given the generally low densities of midges in cocoa plantations.

Figure 7. Banana (*Musa*) pseudostem pieces in various stages of decomposition in this study.

Figure 8. A piece of rotting banana pseudostem with a pocket of cocoa pollination midge larvae.

Table 4. Numbers of cocoa-pollinating midge immatures (larvae and pupae) (Diptera: Ceratopogonidae) in breeding substrates in the field pollination trial (Finca La Tigra, Costa Rica) *Lot B as treatment

Species	Musa pseudostems	leaf litter	total
Lot A			
Forcipomyia cinctipes grp.	--	2	2
Forcipomyia sp.	--	4	4
Total midges (Lot A)	--	6	6
Lot B			
Forcipomyia youngii	270	0	270
F. quatei	1	0	1
F. squamitibia	2	0	2
F. cinctipes group	0	3	3
Forcipomyia sp.	14	0	14
Dasyhelea cacaoi	15	0	15
D. soriai	2	0	2
D. sp. nr.mutabilis	1	0	1
Dasyhelea sp.	20	0	20
Atrichopogon sp.	1	0	1
Total midges (Lot B):	326	3	329

*Data from six different censuses, 1989-1991.

Until October 1991, Lot B has been the experimental plot, receiving the *Musa* pseudostem treatment every six weeks.

Table 5. Numbers of cocoa-pollinating midge immatures (larvae and pupae) (Diptera: Ceratopogonidae) in breeding substrates in the field pollination trial (Finca La Tigra, Costa Rica) with Lot A as treatment*

Species	Musa pseudostems	leaf litter	total
Lot A			
Forcipomyia youngii.	1,543	6	1,549
F. quatei	1	0	1
F. squamitibia	4	0	4
F. genualis	3	0	3
F. sp. 1	1	0	1
F. sp. 2	4	0	4
Lasiohelea sp.	2	0	2
Dasyhelea sp.	49	3	52
Atrichopogon sp.	1	0	1
Lot B			
F. youngii	--	4	4
F. genualis	--	2	2
Dasyhelea sp.	--	1	1
Total midges (Lot A):	1,608	9	1,617

*Data from six different censuses, 1991-1993.

After October 1991, Lot A has been the experimental plot, receiving the *Musa* pseudostem treatment every six weeks.

For the approximately 2500 rotted leaves collected from the control lot (A), a total of 9 pupae of Ceratopogonidae were found, or about 0.04% of the total number of ceratopogonids (326) collected from the rotted pseudostem slices in the experimental lot (B) at the same time. Leaf litter is the most prevalent type of naturally-occurring midge breeding substrate throughout this grove of UF-29 cocoa trees.

It is estimated, on a tentative basis, that the samples of ceratopogonid midges from the experimental lot in particular represent three or four distinct species, including two or three species of *Forcipomyia* (and *Euprojoannisia*) and one or two species of *Dasyhelea*. Cecidomyiid samples most likely include *Mycodiplosis* and *Clinodiplosis*. No larvae or pupae of Cecidomyiidae were found in the litter samples from the control lot. Pseudostems also contained sciarid midge larvae, ants and other small arthropods. Censuses of larval, pupal, and, to a lesser extent, adult stages of cocoa-pollinating midges in banana pseudostem samples (Lot B) and corresponding, approximate quantities of groundcover rotting leaf litter (Lot A) provide a quantitative comparison of the abundance of pollinators between Lots A and B. The samples indicate a clear predominance of ceratopogonid midge species belonging to the *Forcipomyia genualis/quatei/youngii* "group," known pollinators of cocoa in Central America and South America. Samples clearly indicate an elevated abundance of breeding populations of these species in the rotted pseudostems compared to an almost total absence of these midges in the leaf litter samples. Corridors of nearby tropical rainforest are believed to be the source points for midges migrating into the cocoa and "colonizing" the *Musa* pseudostems.

What appears difficult to explain is the prodigious increase in the numbers of *F. youngii* found in the pseudostems when Lot A was the experimental lot (Table 5) as opposed to Lot B. Furthermore, for both lots, there was a tremendous range in the numbers of this midge in the pseudostem pieces. For example, 59 out of 100 pieces had *F. youngii* larvae and pupae when Lot B was the experimental. There was a range of 1-6 immatures per piece but about 90% were found in one piece. In a March 1991 census, a total of 58 immatures of this midge were distributed among 55 pieces of pseudostem. In a February 1993 census three pieces of pseudostem yielded 284 immatures of *F. youngii*. In examining the data closely, it is noteworthy that there is a trend towards higher numbers of midge immatures thriving in the pseudostem pieces in the drier months, i.e., February-March. There was a considerable range in the number of pseudostem pieces examined at each census, due mostly to time constraints. A census period typically lasted 4 or 5 days. For most samples, *F. youngii* larvae and/or pupae occurred in clusters of many individuals; single individuals were rare. When time permitted, observations of cocoa flowers revealed the presence of adult *F. youngii*, especially during morning hours.

Discussion

This study demonstrated that the population of immature stages of cocoa-pollinating midges can be greatly increased through the technique of adding piles of decaying moist pieces of banana pseudostems to the cocoa habitat. However, the cross-over treatment design employing this technique failed to statistically demonstrate a significant increase in cocoa pods with treatment. An earlier non-cross-over study conducted at a different locality in Costa Rica revealed a trend towards increasing pod production by using the banana pseudostem technique (Young 1983). In conducting the present study, it was anticipated that the cross-over experimental design would statistically confirm such a trend, which it did not. While this technique clearly increases the population of immature stages of pollinating midges, it is not known if this translates to a similar increase in the adult numbers. In other words, we do not know that survival rate of midges to the active (pollinative) adult stage.

Our experiment does provide some potential benefit to the cocoa grower, with implications beyond Costa Rica and this study. In terms of the data and analyses, it appears likely that the observed small increase in pod production when banana pseudostems are added to a cocoa plantation has an added benefit to the cacao/cocoa grower. Consider that a typical Central American cocoa plantation tree arrangement is four-by-four meters configuration, about 625 trees per hectare. Our data conservatively suggest that the pseudostem treatment increased pod production by one pod per tree in a plot of about 23 trees. This equates to 26 pods more per hectare or about 3 pounds of dried cocoa beans per hectare. It is our belief, on a preliminary basis, that banana pseudostem treatment, including labor in gathering up and spreading pseudostems in the plantation, does not equate to a dollar value equal or more than the money obtained by selling this increased amount of dried cocoa. Pseudostem treatment, on a tentative basis, appears to be good for growers.

Furthermore, we believe, based upon field data on the numbers of cocoa-pollinating midges found breeding in the pseudostems that increases in pod production results directly from increased availability of pollinators in the plantation. We do not know just how far, on the average, a midge travels in the plantation. This is an important problem because of the relatively close proximity of the two plots. An elevated midge population in the plot with banana pseudostems may lead to more midges moving into the other plot, a distance of only 10-15 meters. This could help explain why the greatest differences in average per tree pod production were found between blocks of trees farthest apart in the two plots.

What is lacking and needed from the study is a set of direct observations on the abundance of adult midges visiting cocoa flowers in both plots, and a confirmation of the frequency of actual pollination events involved. But there can be no doubt from the entomological component of this project that adding pseudostems increases the breeding habitat for key cocoa-pollinating species of midges. We know from hand-pollination studies that (physiologically) cocoa trees can mature greater numbers of pods than what is typically found through natural pollination. Thus, it is reasonable to assume that a greater number of harvestable pods could result from a higher frequency of pollination activity from midges.

If this experiment would be replicated in more seasonal cocoa-growing regions and under drier conditions in general, the results, i.e., enhanced pollination, could be more significant. It is known from Pre-Columbian studies that cocoa was intensely cultivated in drier regions of Central America such as present-day El Salvador (Young 2007 and included references) and where productivity was high. In present-day Central America and elsewhere in the tropics, banana pseudostem pieces kept moist with drip irrigation in dry regions could foster greater cocoa bean production from increased pollinating midge numbers. Cocoa production under such conditions could possibly suppress fungal diseases of the pods. Therefore, a long-term goal could be the diversification of cocoa growing into drier regions of the tropics worldwide. When pollination is a limiting factor in cocoa production, this study could be replicated across an environmental gradient of precipitation levels and seasonality and in which cocoa pod numbers are assessed from very humid to very dry conditions. Such an approach could be taken with the caveat that only highly disturbed areas of habitat would be used and not intact stands of tropical forest. The intent is not to introduce cocoa growing in intact tropical ecosystems, rather to rehabilitate degraded areas with a long-term economically viable cash crop for small-scale growers.

Further studies are needed on the effects of cocoa growing within agroforestry systems. An agroforest would contain a diverse assemblage of shade cover trees whose fruits and seeds are beneficial to both people and wildlife. An agroforest is one means of stabilizing a sustainable income for typically small-scale cocoa growers. Cocoa is a long-lived crop tree which helps to stabilize soil conditions along with other useful shade trees (which can also add income to the cocoa grower). At the same time, agroforests provide food and shelter for migrant bird species as well as nesting sites for resident bird species and other wildlife (e.g., Parrish et al. 2000). Furthermore, cocoa is highly suitable to be planted as a 10-25-year crop in old secondary natural forest that is being selectively logged. While it was recently demonstrated that planting cocoa seedlings within such a habitat greatly suppresses the growth of the trees (Young et. al. 2016), the use of light gaps within secondary wet forest for planting cocoa would be more promising.

In conclusion, although the result of the field experiment summarized in this chapter did not yield significant response to treatment, the technique of using moist rotting banana pseudostems as cocoa-pollination midge breeding sites to raise levels of pollinators has potential for success under different conditions as discussed above. A return to perhaps growing cocoa in moist, shaded gullies in dry regions and enhancing moisture levels via drip irrigation could elicit a powerful response for pollinating midges to proliferate and increase pod yields.

We also propose that cocoa trees provide a diversified environmental benefit to both people and nature. For example, at the Tirimbina Rainforest Center in northeastern Costa Rica, eco-tourists can participate in a "Chocolate Tour." Being one of the most popular commodities in the world, a guided tour within a tropical rainforest setting provides an opportunity for eco-tourists to learn about chocolate, from cocoa bean to bar. This educational program includes participants gathering cocoa beans from split-open pods, drying and roasting beans and making mini chocolate bars in the shape of rainforest animals. Emphasis is also placed on the relation of the cocoa tree to the rainforest ecosystem.

From an environmental standpoint, cocoa has several diverse benefits, from being an eco-friendly sustainable cash crop for small-scale growers, ideal candidate for agroforestry systems and managed secondary rainforest, promotion of wildlife corridor habitat, to having a strong eco-tourism potential. All of these benefits derive ultimately from effective levels of midge-mediated pollination of cocoa flowers.

ACKNOWLEDGMENTS

The research presented here was funded by grants from the American Cocoa Research Institute of the Chocolate Manufacturers Association of the United States of America. We thank Laura Deegan and Raymond Guries of the Department of Forestry at the University of Wisconsin-Madison for assistance with data analysis and helpful discussion of the project. Field assistance was provided by Lucrecia Guillen, Charles Hunter and Jorge Mejias. We thank the Milwaukee Public Museum and the University of Wisconsin for general support. We thank Nikki Hammerberg for her excellent assistance in the final preparation of the manuscript.

REFERENCES

Cope, F. W. 1939. Agents of pollination in cacao. *Annual Report on Cacao Research* (Trinidad), 1939: 13-19.

Dessart, P. 1961. Contribution a l'etude des Ceratopogonidae (Diptera). *Bulletin Agriculturique du Congo Belge* [Contribution to the study of Ceratopogonidae (Diptera). *Agricultural Newsletter of the Belgian Congo*] 52: 525-540.

Parrish, J. R.; Reitsma, R.; Greenberg, R.; Skerl, K., McLarner, W., Mack, R. & Lynch, J. 2000. *Cacao as a crop and conservation tool in Latin America: Addressing the needs of farmers and forest biodiversity.* Arlington, VA: The Nature Conservancy.

Soria, S. de J. 1970. *Studies on Forcipomyia spp. midges (Diptera, Ceratopogonidae) related to the pollination of Theobroma cacao L.* PhD diss. University of Wisconsin-Madison.

Soria, S. de J. & Wirth, W. W. 1979. Ceratopogonid midges (Diptera: Nematocera) collected from cacao flowers in Palmira, Colombia. *Revista Theobroma* 9: 77-84.

Wellensiek, S. J. 1932. *Flower-biological observations with cocoa.* Indonesie 6: 87-101. Archive Koffeicult (Dutch translation).

Winder, J. A. & Silva, P. 1975. Current research on insect pollination of cocoa in Bahia. In *Proceedings of the Fourth International Cocoa Research Conference*, Trinidad & Tobago, 553-565.

Young, A. M 1982. Effects of shade cover and availability of midge breeding sites on pollinating midge populations and fruit-set in two cocoa farms. *Journal of Applied Ecology* 19: 47-63.

Young, A. M. 1983. Seasonal differences in abundance and distribution of cocoa pollinating midges in relation to flowering and fruit-set between sunny and shaded habitats of the La Lola Cocoa Farm in Costa Rica. *Journal of Applied Ecology* 20: 801-831.

Young, A. M. 1986. Habitat differences in cocoa tree flowering, fruit-set and pollinator availability in Costa Rica. *Journal of Tropical Ecology* 2: 163-186.

Young, A. M. 2007. *The Chocolate Tree – A Natural History of Cacao.* University Press of Florida, 219 pp.

Young, A. M. 2017. *Sarapiqui Chronicle – A Naturalist in Costa Rica.* Second Edition. University of New Mexico Press, 350 pp.

Young, A. M., Barry, K. & Schnitizer, S. A. 2016. Top-down versus bottom-up ecological control in cacao, *Theobroma cacao* L. Tropical Agriculture (Trinidad) 93: 216-226.

In: The Diversified Benefits of Cocoa … ISBN: 978-1-53613-258-8
Editor: Bonifacia Zayas Espinal

Chapter 6

ANALYSING THE DEPENDENCE BETWEEN COCOA SOLIDS IN CHOCOLATE AND THE CONTENT OF POLYPHENOLS, MINERALS AND DIETARY FIBER

Ivana Lončarević[1,*], Biljana Pajin[1], Aleksandra Torbica[2], Vesna Tumbas Šaponjac[1], Jovana Petrović[1] and Danica Zarić[3]
[1]Faculty of Technology, University of Novi Sad, Novi Sad, Serbia
[2]Institute of Food Technology, University of Novi Sad, Novi Sad, Serbia
[3]IHIS Techno experts d.o.o., Research Development Center, Beograd, Srbija

ABSTRACT

The raw materials used for chocolate production provide a source of proteins, carbohydrates, fats, minerals, and vitamins, which are essential

---

* Corresponding author: ivana.radujko@tf.uns.ac.rs.

for growth and development. Dark chocolate is also rich in polyphenolic compounds, originating from dark solids of cocoa beans. Milk chocolate contains less cocoa bean solids and thus less polyphenolic compounds comparing to dark chocolate. On the other hand, white chocolate contains only cocoa butter and thus lack bioactive components that have a positive impact on human health. Cocoa bean contains several minerals, some of which are found in high amounts in processed chocolate. The amount of retained minerals depends on the content of cocoa bean solids in chocolate. Accordingly, dark chocolate typically has a higher amount of minerals than milk or white chocolate. It is also known that unprocessed cocoa bean presents a good source of dietary fiber, mostly insoluble fiber, which is significantly reduced when removing the cocoa husk. Dark chocolate does not contribute significantly to dietary fiber intake, but, however, contains a higher amount of dietary fiber compared to milk chocolate.

This research examined and compared the content of total polyphenols, minerals (potassium, calcium, magnesium, iron, zinc, copper, manganese), total dietary fiber and soluble dietary fiber in in different chocolate products collected from the market: white chocolate (25% of cocoa solids – cocoa butter), milk chocolate (25% of cocoa solids), baking chocolate containing 44% of cocoa solids, and dark chocolates containing 58%, 75%, 88%, and 100% of cocoa solids.

INTRODUCTION

Chocolate is a complex multiphase system often defined as a suspension consisting of non-fat particles (sugar, cocoa solids and milk powder) dispersed in continuous fat phase (cocoa butter and milk fat) (Zarić et al., 2016). Primary chocolate categories are dark, milk and white that differs in the content of cocoa solids (fat free cocoa solids and cocoa butter) and milk fat (Rossini et al., 2011). Dark chocolate is formulated with cocoa liquor, cocoa butter, sugar, and lecithin, where in the case of milk chocolate, milk fat is added. On the other hand, in white chocolate, the cocoa is left out completely (Samsudin, 1996).

Although enjoyed by many for its flavor and textural properties, chocolate is typically viewed as a non-nutritional confectionery food that provides only fats and calories. However, certain chocolates can provide significant amounts of a number of essential nutrients (Steinberg et al.,

2003). The cocoa bean, the raw material for chocolate production, has two important types of cells within the cotyledons: storage cells containing fat and proteins, and the pigment cells containing polyphenolic compounds and methylxanthines (theobromine and caffeine). The total content of phenolic compounds is approximately 6-8% and can contain as much as 10% based on the dry matter of the bean, which depends on the variety of cocoa beans used for mass production. It has been shown that processes such as fermentation, drying, alkalization, and roasting contribute to the degradation of phenolic compounds. However, with the proper processing and manufacturing controls in place, substantial amounts of these potentially beneficial compounds can be retained from the cocoa bean (Steinberg et al., 2003; Fowler, 2009; Żyżelewicz et al., 2016[a]). Therefore, much attention has been paid to cocoa and cocoa products due to their potential positive impact on cardiovascular health, mostly contributed by the polyphenolic fraction of cocoa (Lecumberri et al., 2007). Although numerous polyphenolic compounds are present in the cocoa bean, cocoa powder is particularly rich in a subclass of polyphenolics known as flavonoids. The primary flavonoids in cocoa and chocolate are the flavan-3-ols catechin and epicatechin (monomeric units) and proanthocyanidins (also termed procyanidins), which are polymeric compounds comprising catechin and epicatechin subunits (Steinberg et al., 2003). Cocoa-rich products have more flavonoids and greater antioxidant capacity per serving than tea, red wine, fruits, and vegetables, which are renowned for their high content of flavonoids (Shiina et al., 2009). Flavonoids influence cocoa quality, as they impart astringent flavors and drive a wide range of colors, from violet to dark brown. The bioactivity of the cocoa flavonoids include antioxidant, anti-inflammatory, immune modulating, and antimicrobial activities (Mayorga-Gross et al., 2016). Given that dark chocolate is formulated with a higher percentage of cocoa bean liquor than milk chocolate, it often contains greater amounts of flavonoids. This is an important distinction because not all chocolates are equal sources of flavonoids (Steinberg et al., 2003). White chocolate does not contain dark cocoa particles but only cocoa butter, sugar, milk, and lecithin. Thus, white chocolate lacks cocoa polyphenols (Rimbach et al., 2011).

Cocoa contains many essential minerals including iron, copper, zinc, magnesium, phosphorus, manganese and potassium, all of which have important roles in the metabolism of the human body. The amounts vary and are dependent on the soil in which the cocoa is grown. The final mineral content of chocolate is also dependent on the proportions of cocoa liquor and milk in the final product, where high cocoa solid chocolate is rich in iron, copper, and manganese. Milk chocolate also provides important nutritional minerals since milk is an excellent source of calcium and potassium, then sodium, phosphorus and magnesium (Lambert, 2009). Mineral content in cocoa products also varies depending on the extent and type of processing (Steinberg et al., 2003).

Besides being an excellent source of polyphenols and minerals, cocoa liquor as raw material for chocolate production contains 17.5% of total dietary fiber (14% of insoluble fiber and 3.5% of soluble fiber) (Pajin, 2014).

Taking into account the significance of bioactive components in chocolate, this chapter examined and compared the content of total polyphenols, minerals (potassium, magnesium, calcium, iron, zinc, manganese, copper), and dietary fiber in chocolates with different content of cocoa solids.

Materials and Methods

Starting Materials

Material used in this chapter included chocolates collected from the market: white chocolate (25% of cocoa butter, in the following text marked as W25), milk chocolate (25% of cocoa solids, M25), baking chocolate (44% of cocoa solids, B44), and dark chocolates containing 58% (D58), 75% (D75), 88% (D88), and 100% (D100) of cocoa solids.

Methods

Determination of Total Polyphenol Content (TPC)

For determination of total polyphenol content in chocolate samples the extraction procedure according to the Belščak-Cvitanović et al. (2012) was applied. In order to eliminate lipids from the samples, 2.0 g of chocolate sample was extracted three times with 10 mL of n-hexane. The defatted cocoa solids were air-dried for 24 h to remove the residual organic solvent (Adamson et al., 1999). The phenolic compounds were extracted two times from 2.0 g of defatted chocolate samples with 10 mL of aqueous methanol (70%, v/v) for 30 min in an ultrasonic bath Belščak-Cvitanović et al. (2015). After each extraction, the mixture was centrifuged for 10 min at 3000 rpm and the supernatant was decanted. The supernatants were combined and filtered to remove the residual particles. The total polyphenol contents in samples were determined spectrophotometrically by a Folin-Ciocalteau method adapted to microscale (González-Molina et al., 2008). Briefly, the reaction mixture was prepared in 96 well microplate by mixing 15 μL of extract, 170 μL of distilled water, 12 μL of the Folin-Ciocalteu's reagent (2 M) and 30 μL of 20% sodium carbonate. After 1 h, the absorbance at 750 nm was obtained using distilled water as a blank. Results were expressed as gallic acid equivalents (GAE) per 100 g of sample.

Determination of Mineral Composition

Determination of mineral composition was performed on atomic absorption spectrophotometer SpectrAA-10 (Varian, Australia) according to SRPS EN ISO 6869: 2008 method.

Determination of Dietary Fiber

Dietary fiber in chocolate samples was determined according to the A.O.A.C (2000) method.

Results and Discussion

Total Polyphenol Content

Total polyphenol content in white, milk and dark chocolates is presented in Table 1.

As expected, white chocolate lacks cocoa polyphenols, containing only 29.64 mg GAE/100 g. Mursu et al. (2004) reported the total amount of catechins in white chocolate to be 0.3 mg/100 g with no procyanidins present. The addition of certain amount of dark cocoa particles in milk chocolate with 25% of cocoa solids increased total polyphenol content to 318.48 mg GAE/100 g. Miller et al. (2006) reported total polyphenol content in different milk chocolates ranging from 325 mg GAE/100 g to 538 mg GAE/100 g. The increase of cocoa solids from 25% to 44% and 58% influences significant increase of total polyphenol content from 318.48 mg GAE/100 g to 978 and 1741.98 mg GAE/100 g, respectively. On the other hand, the increasing cocoa solids in chocolate from 58% to 75% and 88% increases the content of total polyphenols, but lower than expected (from 1741.98 mg GAE/100 g to 1852.72 and 1912.49 mg GAE/100 g, respectively). However, chocolate made of 100% cocoa solids contains 2823.76 mg GAE/100 g. The content of total polyphenols in chocolates does not increase proportionally with an increase of cocoa solids since chocolates are taken from the market and do not have the same origin of cocoa. Natsume et al. (2000) analysed polyphenols in cocoa liquor from six different countries. The results showed almost the same proportions of the proanthocyanidin components in each of them, but the quantity of each proanthocyanidin component differed markedly. They reported that this type of difference might be due to variations in the type of beans used or differences in the fermentation process.

Table 1. Total polyphenol content in chocolates with different amount of cocoa solids

Sample	Total polyphenol content (mg GAE/100 g)
W25	29.61 ± 1.48
M25	318.48 ± 6.24
B44	978.68 ± 45.61
D58	1741.98 ± 75.53
D75	1852.72 ± 40.75
D88	1912.49 ± 81.62
D100	2823.76 ± 71.42

Results represent the mean, n=3.

Mineral Composition

The content of potassium, magnesium, calcium, and iron in examined chocolates is presented in Figure 1.

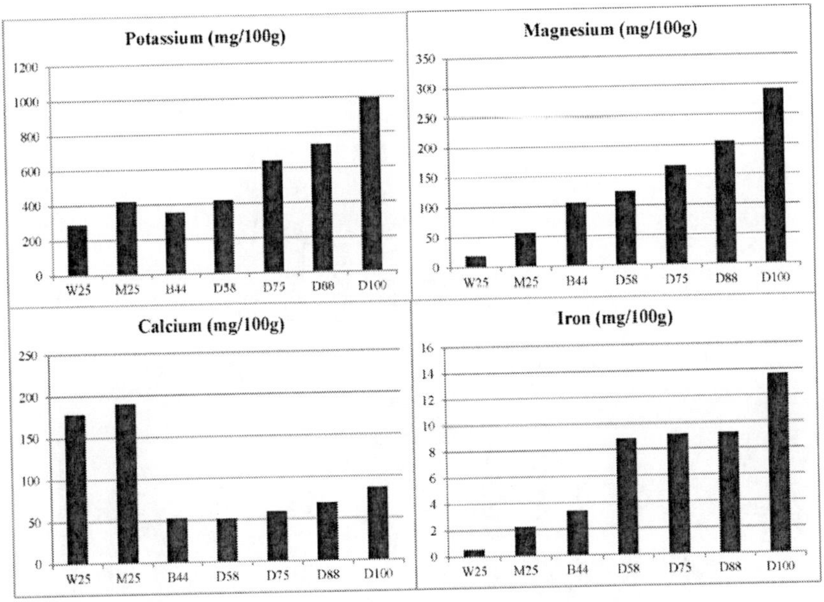

Figure 1. Potassium, magnesium, calcium, and iron content in different types of chocolate.

Dark chocolate, made of 100% of cocoa solids, is the richest source of potassium, magnesium, iron, zinc, manganese, and copper compared to all examined chocolates. Also, it contains the highest amount of calcium in relation to other dark chocolate samples (86.2 mg/100 g). On the other hand, milk chocolate is the richest source of calcium (190.52 mg/100 g), while white chocolate also contains an initial quantity of this mineral (178.51 mg/100 g). Furthermore, milk chocolate has a higher content of potassium than baking chocolate and dark chocolate with 58% of cocoa solids. Nevertheless, dark cocoa parts significantly increase the content of potassium in chocolates with 75, 88, and 100% of cocoa solids where D100 has more than the double higher value of potassium comparing to D58. Milk powder enriched white chocolate with magnesium in a quantity of 19.13 mg/100 g. However, all other samples have a higher amount of this mineral where the content of this mineral increases with the increase of dark cocoa solids and amounts 291.6 mg/100 g in chocolate with 100% of cocoa solids. The content of iron also increases with the increase of dark cocoa solids, amounting 0.52 mg/100 g in W25 to 13.64 mg/100 g in D100.

The content of zinc, manganese, and copper is presented in Figure 2.

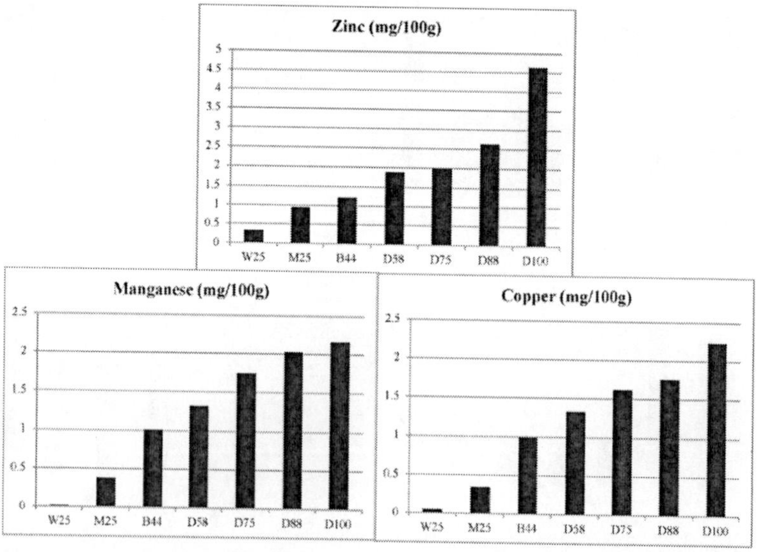

Figure 2. Zinc, manganese, and copper content in different types of chocolate.

White chocolate is a poor source of zinc, manganese, and copper. Dark cocoa particles remarkably increase their content already in chocolate with 25% of cocoa solids. Further increasing of cocoa solids in chocolates leads to an increase of those minerals, where D100 is the richest source of zinc, manganese, and copper (4.65 mg/100 g, 2.17 g/100 g, and 2.24 g/100 g, respectively).

Dietary Fiber (DF) Content

The content of total, insoluble and soluble dietary fiber (DF) in examined chocolate samples is presented in Table 2.

Table 2. Dietary fiber in white, milk and dark chocolate

Sample	Total DF (%)	Insoluble DF (%)
W25	n.d.*	n.d.*
M25	2.18 ± 0.51	1.72 ± 0.33
B44	6.44 ± 0.74	5.93 ± 0.27
D58	8.11 ± 0.63	7.91 ±0.63
D75	9.26 ± 0.88	8.59 ± 0.54
D88	10.60 ± 0.72	10.00 ± 0.69
D100	17.12 ± 0.55	14.63 ± 1.09

Results represent the mean, n=3.* n.d. – not detected

Dietary fiber was not detected in white chocolate due to the lack of dark cocoa solids. At the same time, milk chocolate with the same content of cocoa solids, but with a certain amount of dark cocoa solids in addition to cocoa butter, contains 2.18% of total DF, of which 1.72% insoluble. The increase of cocoa solids in chocolate led to an increase in the content of DF, but not proportionally. Dietary fiber in all chocolates is mostly made up of soluble dietary fiber. Dark chocolate contains from 8.11% of total DF (in D58) to 9.26% in D75 and 10.60% in D88. Dark chocolate with 100% of cocoa solids contains the highest share (17.12%) of total DF, of which 14.63 insoluble DF, compared to all other chocolate samples.

Conclusion

Although chocolate is often viewed as a food with minimal nutritional value, some types of chocolate contain a remarkable amount of polyphenol components and minerals. As expected, white chocolate lacks polyphenols and contains the lowest share of potassium, magnesium, iron, zinc, manganese, and copper. On the other hand, a significant amount of calcium is provided in white chocolate by milk powder. Milk chocolate contains the highest share of calcium compared to all other examined chocolates. Milk powder is also a good source of potassium, where milk chocolate with 25% of cocoa solids contains higher content of potassium than baking chocolate and dark chocolate with 58% of cocoa solids. The share of cocoa solids in different types of chocolate has a direct impact on the content of magnesium, iron, zinc, manganese, and copper where chocolate with 100% of cocoa solids has the highest content of those minerals.

Milk and dark chocolates also contain dietary fiber originating from dark cocoa solids. Although chocolate consumption does not contribute significantly to dietary fiber intake, it must be emphasized that the share of dark cocoa solids in chocolate has a direct contribution to dietary fiber content with the amount of 17.12% in chocolate made of 100% cocoa solids.

Acknowledgments

This research has been supported by Provincial Secretariat for Higher Education and Scientific Research, Autonomous Province of Vojvodina, Republic of Serbia (Project no. 142-451-3586/2016-03) and the Ministry of Science and Technological Development of the Republic of Serbia (Project no. 31014).

References

Adamson, G.E., Lazarus, S.A., Mitchell, A.E., Prior, R.L., Cao, G., Jacobs, P.H. (1999). HPLC method for the quantification of procyanidins in cocoa and chocolate samples and correlation to total antioxidant capacity, *Journal of Agricultural and Food Chemistry,* 47, 4184-4188.

AOAC (2000). *Official methods of analysis of the association of official analytical chemists international.* Vols. I and II. 17th ed. Arlington: Association of Official Analytical Chemists Inc.

Belščak-Cvitanović, A., Komes, D., Benković, M., Karlović, S., Hečimović, I., Ježek, D., Bauman, I. (2012). Innovative formulations of chocolates enriched with plant polyphenols from *Rubus idaeus L.* leaves and characterization of their physical, bioactive and sensory properties, *Food Research International,* 48, 820 830.

Belščak-Cvitanović, A., Komes, D., Dujmović, M., Karlović, S., Biškić, M., Brnčić, M., Ježek, D. (2015). *Physical, bioactive and sensory quality parameters of reduced sugar chocolates formulated with natural sweeteners as sucrose alternatives,* 167, 61-70.

Đudarić, L., Fužinac-Smojver, A., Muhvić, D., Giacometti, J. (2015). The role of polyphenols on bone metabolism in osteoporosis, *Food Research International,* 77, 290-298.

Fowler, M.S. (2009). Cocoa beans: from tree to factory. In: S.T. Beckett (Ed.), *Industrial chocolate manufacture and use* (pp. 21), Blackwell Publishing Ltd., York, UK.

González-Molina, E., Moreno, D.A., García-Viguera, C. (2008). Genotype and harvest time influence the phytochemical quality of Fino lemon juice (*Citrus limon* (L.) Burm. F.) for industrial use, *Journal of Agricultural and Food Chemistry,* 56, 1669–1675.

ISO 6869 (2000). *Animal feeding stuffs - Determination of the contents of calcium, copper, iron, magnesium, manganese, potassium, sodium and zinc - Method using atomic absorption spectrometry.*

Lambert, J.P. (2009). Nutrition and health aspects of chocolate. In: S.T. Beckett (Ed.), *Industrial chocolate manufacture and use* (pp. 21), Blackwell Publishing Ltd., York, UK.

Lecumberri, E., Goya, L., Mateos, R., Alía, M., Ramos, S., Izquierdo-Pulido, M., Bravo, M. (2007). A diet rich in dietary fiber from cocoa improves lipid profile and reduces malondialdehyde in hypercholesterolemic rats. *Nutrition,* 23, 332-341.

Mayorga-Gross, A.L., Quirós-Guerrero, L.M, Fourny, G., Vaillant, F. (2016). An untargeted metabolomic assessment of cocoa beans during fermentation. *Food Research International,* 89, 901-909.

Miller, K.B., Stuart, D.A., Smith, N.L., Lee, C.Y., Mchale, N.L., Flanagan, J.A., Ou, B., Hurst, W.J., (2006). Antioxidant Activity and Polyphenol and Procyanidin Contents of Selected Commercially Available Cocoa-Containing and Chocolate Products in the United States. *Journal of Agricultural and Food Chemistry,* 54, 4062-4068.

Mursu, J., Voutilainen, S., Nurmi, T., Rissanen, Virtanen, J.K. Kaikkonen, J., Nyyssonen, K., Salonen, J.T. (2004). Dark chocolate consumption increases HDL cholesterol concentration and chocolate fatty acids may inhibit lipid peroxidation in healthy humans. *Free Radical Biology & Medicine,* 37, 1351–1359.

Pajin, B. (2014). *The technology of chocolate and related cocoa products.* Faculty of Technology, University of Novi Sad, Novi Sad, Serbia (Chapter 2, pp. 77).

Natsume, M., Osakabe, N., Yamagishi, M., Takizawa, T., Nakamura, T., Miyatake, H., Hatano, T., Yoshida, T. (2000). Analyses of polyphenols in cocoa liquor, cocoa, and chocolate by normal-phase and reversed-phase HPLC. *Bioscience, Biotechnology, and Biochemistry,* 64, 2581-2587.

Pedan, V., Fischer, N., Rohn, S. (2016). An online NP-HPLC-DPPH method for the determination of the antioxidant activity of condensed polyphenols in cocoa. *Food Research International,* 89, 890-900.

Rimbach, G., Egert S. & de Pascual-Teresa, S. (2011). Chocolate: (un)healthy source of polyphenols, *Genes & Nutrition,* 6, 1-3.

Rossini, K., Noreña C.P.Z., Brandelli, A. (2011). Changes in the color of white chocolate during storage: potential roles of lipid oxidation and non-enzymatic browning reactions, *Journal of Food Science and Technology,* 48, 305–311.

Samsudin, S. (1996). Use of Palm Mid-Fraction in White Chocolate Formulation, *Journal of the Science of Food and Agriculture,* 71, 483-490.

Shiina, Y, Funabashi, N., Lee, K., Murayama, T., Nakamura, K., Wakatsuki, Y., Daimon, M., Komuro, I. (2009). Acute effect of oral flavonoid-rich dark chocolate intake on coronary circulation, as compared with non-flavonoid white chocolate, by transthoracic Doppler echocardiography in healthy adults, *International Journal of Cardiology,* 31, 424-429.

Steinberg, F.M., Bearden M.M., Keen, C.L. (2003). Cocoa and chocolate flavonoids: Implications for cardiovascular health. *Journal of the American Dietetic Association*, 103, 215-223.

Zarić, D., Bulatović, M., Rakin, M., Krunić, T., Lončarević, I., Pajin, B. (2016). Functional, rheological and sensory properties of probiotic milk chocolate produced in a ball mill, *RSC Advances,* 6, 13934–13941.

Żyżelewicz, D., Krysiak, W., Oracz, J., Sosnowska, D., Budryn, G., Nebesny, E. (2016)[a]: The influence of the roasting process conditions on the polyphenol content in cocoa beans, nibs and chocolates. *Food Research International,* 89, 918-929.

Żyżelewicz, D., Zaklos-Szyda, M., Juśkiewicz, J., Bojczuk, M., Oracz, J., Budryn, G., Miśkiewicz, K., Krysiak, W., Zduńczyk, Z., Jurgoński, A. (2016)[b]. Cocoa bean (Theobroma cocoa L.) phenolic extracts as PTP1B inhibitors, hepatic HepG2 and pancreatic βTC3 cell cytoprotective agents and their influence on oxidative stress in rats. *Food Research International,* 89, 946-957.

INDEX

D

E

F

G

H

I

K

L

M

N

O

P

Q

R

S

T

U

V

W

Z